D0983109

Mechanical Engineering

Sample Examination

Third Edition

Michael R. Lindeburg, PE

Professional Publications, Inc. • Belmont, CA

Production Manager: Aline Sullivan Magee
Acquisitions Editor: Tom Tolfa
Project Editor: Mia Laurence
Copy Editor: Mia Laurence
Book Designer: Charles P. Oey
Typesetter: Cathy Schrott
Illustrator: Yvonne M. Sartain
Proofreader: Margaret S. Yoon
Cover Designer: Charles P. Oey

MECHANICAL ENGINEERING SAMPLE EXAMINATION
Third Edition

Copyright © 1998 by Professional Publications, Inc. All rights reserved. No part of this publication may be reproduced, stored in a retrieval system, or transmitted, in any form or by any means, electronic, mechanical, photocopying, recording, or otherwise, without the prior written permission of the publisher.

Printed in the United States of America

Professional Publications, Inc.
1250 Fifth Avenue, Belmont, CA 94002
(650) 593-9119
www.ppi2pass.com

Current printing of this edition: 1

Library of Congress Cataloging-in-Publication Data
Lindeburg, Michael R.
    Mechanical engineering sample exam / Michael R. Lindeburg.--
3rd ed.
        p.   cm.
    ISBN 1-888577-17-7
    1. Engineering--United States--Examinations--Study guides.
2. Mechanical engineering--United States--Examinations--Study
guides. 3. Engineering--Problems, exercises, etc. 4. Engineers--
Certification--United States. I. Title.
TA159.L575   1997
621'.076--dc21   97-31464
                                                        CIP

Preface to the Third Edition

Although the types of problems in the PE licensing exam for mechanical engineers have remained essentially unchanged for many years, the format of the exam continues to evolve. This edition of the *Sample Examination* reflects NCEES' reduction from five answer choices to four in the multiple-choice problems. It also reflects the placement of all free-response problems in the morning session, while all multiple-choice problems are in the afternoon session.

The number of problems in each exam topic is consistent with the NCEES subject breakdown. The most noticeable change is the inclusion of two new HVAC problems, a subject that was not represented by any problems in the previous edition.

Even though engineering economics is no longer a formal exam topic, elements of that subject continue to appear in various problems. You will find one "real" engineering problem in this sample exam disguised as an engineering economics problem (or, vice versa!).

Approximately two-thirds of the problems in this edition are new. As with most of the problems that I now write for any of my exam review publications, the inspiration for these new problems has come from actual examinees who made suggestions on how sample exam coverage could be enhanced.

Typesetting and illustrating in this edition now follow Professional Publications' strict style guide for engineering publications. I am sure you will benefit greatly from the flow, completeness, and appearance of the problem statements and solutions.

Those awful block-out screens (that nobody liked) have been omitted from this edition. Once again, you'll be able to make notes in your book.

There is very little use of SI (metric) units in this sample exam. This is consistent with the current mechanical PE exam. In comparison to the FE (Fundamentals of Engineering), which essentially uses only SI units, the mechanical PE exam continues to feature predominantly English units.

Although full solutions are provided, none of the "criteria" that NCEES uses for grading are included. The NCEES grading methodology is now well known and is fully described in the *Mechanical Engineering Reference Manual*, so it doesn't need to be illustrated in this publication.

This is the first of my supplementary mechanical publications to be revised following the release of the 10th edition of the *Mechanical Engineering Reference Manual*. Therefore, this new edition is consistent in nomenclature and style with the 10th edition of the *Reference Manual*.

As in all of my publications, I invite your comments. If you disagree with a solution, or if you think there is a better way to do something, please let me know. You can use the business reply cards at the back of this book to contact me.

Thank you for buying another one of my books!

Michael R. Lindeburg, PE
Belmont, CA

How to Use this Book

This book is a sample exam, so there are only a few ways that you can use it. Some people will work through every problem, basically using it as a collection of solved problems. At the other extreme, some people will run out of time and won't use this at all. But, to me, the main issue is not *how* you use this sample exam, but *when* you use it.

Though I tried to include realistic exam problems, I did not write this book intending it to be a diagnostic tool guiding your preparation. You shouldn't take this sample exam and then design your review around what you didn't know. If you take this exam and don't do well on a particular problem, I wouldn't want you to spend the next three months preparing for that type of problem. The tried-and-true method of exam preparation is a systematic, thorough, and complete approach based on long-term exam trends, not based on transient and oddball fads. The *Mechanical Engineering Reference Manual* is what you should use for extensive preparation.

The value of a sample exam does not lie in its ability to guide your preparation. Rather, the value is in giving you an opportunity to bring together all of your knowledge and to practice your test-taking skills. The three most important skills are (1) selecting the right problems, (2) organizing your references and other resources, and (3) managing your time. I intended this sample exam to be taken within a few weeks of your actual exam. That's the only time that you will be able to focus on test-taking skills without the distraction of rusty recall.

You'll need to set aside an entire day to take this sample exam the way I intended it to be used. I know that using up another day is asking a lot from you. But if you start early enough and study diligently, by the time the actual exam rolls around, you will probably be weak in only one area: familiarity with the nature of the exam.

In athletics, coaches often speak of the home-field advantage. Athletes who are comfortable in their environment play better. Well, examinees who have "seen it before" via a sample exam have a psychological edge, as well. This publication was written to give that edge to you.

Good luck!

The National Society of Professional Engineers

Whether you design water works, consumer goods, or aerospace vehicles; whether you work in private industry, for the U.S. government, or for the public; and whether your efforts are theoretical or practical, you (as an engineer) have a significant responsibility.

Engineers of all types perform exciting and rewarding work, often stretching new technologies to their limits. But those limits are often incomprehensible to nonengineers. As the ambient level of technical sophistication increases, the public has come to depend increasingly and unhesitatingly more on engineers. That is where professional licensing and the National Society of Professional Engineers (NSPE) become important.

NSPE, the leading organization for licensed engineering professionals, is dedicated to serving the engineering profession by supporting activities such as continuing educational programs for its members, lobbying and legislative efforts on local and national levels, and the promotion of guidelines for ethical service. From local, community-based projects to encourage top-scoring high school students to choose engineering as a career

to hard-hitting lobbying efforts in the nation's capital to satisfy the needs of all engineers, NSPE is committed to you and your profession.

Engineering licensing is a two-way street: It benefits you while it benefits the public and the profession. For you, licensing offers a variety of benefits, ranging from peer recognition to greater advancement and career opportunities. For the profession, licensing establishes a common credential by which all engineers can be compared. For the public, a professional engineering license is an assurance of a recognizable standard of competence.

NSPE has always been a strong advocate of engineering licensing and a supporter of the profession. Professional Publications hopes you will consider membership in NSPE as the next logical step in your career advancement. For more information regarding membership, write to the National Society of Professional Engineers, Information Center, 1420 King Street, Alexandria, VA 22314, or call (703) 684-2800.

Notice to Examinees

Do not copy, memorize, or distribute problems from the Principles and Practice of Engineering (P&P) Examination. These acts are considered to be exam subversion.

The P&P examination is copyrighted by the National Council of Examiners for Engineering and Surveying. Copying and reproducing P&P exam problems for commercial purposes is a violation of federal copyright law. Reporting examination problems to other examinees invalidates the examination process and threatens the health and welfare of the public.

Instructions

Name: _____
　　　　 Last　　　　 First　　　 Middle Initial

You may use textbooks, handbooks, bound reference materials, and any battery-powered, silent calculators to work this examination. However, no blank papers, writing tablets, unbound tables, or unbound notes are permitted. Sufficient room for scratch work is provided in the Solution Pamphlet. You are not permitted to share or exchange materials with other examinees.

You will have four hours in which to work each part of the test. Your score will be determined by the number of problems (up to four) that you solve correctly. Each correct solution is worth ten points. The maximum possible score for the A.M. part of the examination is 40 points. Except for the objective problems (i.e., multiple choice format), partial credit for partially correct solutions will be given.

Do not submit solutions or partial solutions for more than four problems. Do not enter solutions into this Examination Booklet. All solutions must be entered in the special Solution Pamphlet supplied by your proctor.

Record your answers to the objective problems with a No. 2 pencil on the special machine-readable pages that make up the first part of your Solution Pamphlet. No credit will be given for objective problem solutions that are not written on these pages.

If you finish early, check your work and make sure you have followed all instructions. After checking your solutions, you may turn in your Examination Booklet and Solution Pamphlet and leave the examination room. Once you leave, you will not be permitted to return to work on your solutions.

Do not work any problems from the PART 2—P.M. test during the first four hours of this test.

Principles and Practice of Engineering Examination

Sample Examination

Part 1—A.M.

1 SITUATION

An electrical generator and an air conditioning unit on a passenger aircraft use 68 lbm/min of compressed air at 100 psia and 640°F bled off from one of the jet engine compressor sections. The air conditioner runs on an open cycle and maintains the passenger cabin at 80°F and 12 psia. The electrical generator is driven by an air turbine whose expansion is a polytropic process with a polytropic exponent of 1.2. The mechanical efficiency of the turbine is 85%. Prior to entering the air turbine, compressed air from the engine is cooled to 250°F in a crossflow heat exchanger using ram air from outside the aircraft. The pressure drop across the heat exchanger is negligible.

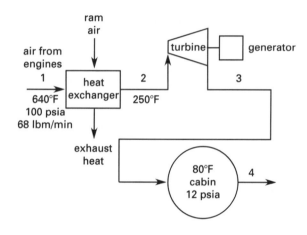

REQUIREMENT

(a) Assuming that air is a real gas, what power (in kilowatts) is developed by the turbine?

(b) Assuming that air is a real gas, what is the total cooling load from the passenger cabin?

(c) Repeat part (a) assuming air is an ideal gas, the turbine expansion is isentropic, and the mechanical efficiency remains the same.

(d) Repeat part (b) assuming air is an ideal gas, the turbine expansion is isentropic, and the mechanical efficiency remains the same.

2 SITUATION

A supply of 150 psia, 500°F steam is available to drive a low-pressure steam turbine. The turbine exhausts to a 2 psia condenser. The isentropic efficiency of the turbine is 86%. The turbine loading is constant, but it does not require the entire energy content of the steam. Therefore, the steam pressure is currently being reduced from 150 psia to 100 psia in an ideal throttling process prior to entering the turbine. To conserve energy, it is proposed that a fraction of the steam be bled off for heating feedwater. Then, the pressure-reducing throttle valve can be opened fully without changing the turbine work. It is proposed that 30 psia steam be bled off. Feedwater currently enters the steam generator at 60 psia and 60°F. The isentropic efficiency for the bleed fraction is assumed to be the same as for the full expansion.

REQUIREMENT

(a) What is the maximum percentage of the steam flow that can be bled off without reducing the turbine work output?

(b) What is the maximum temperature to which the feedwater can be heated?

3 SITUATION

A building is located where the outside conditions are 96°F dry bulb and 76°F wet bulb. The building is air conditioned and all rooms are to be maintained at 76°F dry bulb and 50% relative humidity. The instantaneous total heat gain from all sources (including conduction and solar) is 150,000 Btu/hr, and the sensible load is 80% of the total load. 800 ft^3/min of ventilation air are brought in from the outside. The outside air is conditioned to 58°F dry bulb before entering the conditioned space.

REQUIREMENT

(a) What air flow rate through the building is required such that all internal heat gain is removed?

(b) What is the humidity (in grains per pound) of the air supplied to the conditioned space?

(c) What is the minimum capacity (in tons) of the air conditioner?

4 SITUATION

A 2 in diameter solid shaft made from hardened AISI-1020 cold-drawn steel is supported on two ball bearings. The shaft carries two 20 in diameter massless pulleys, one of which is the driving pulley and the other which is the driven pulley. The rotational speed of the shaft is much less than its critical speed. The shaft's yield and ultimate tensile strengths are 48 ksi and 69 ksi, respectively. The endurance limit in complete reversal is 35 ksi. Poisson's ratio is 0.283. The maximum lateral deflection of the shaft at any point is limited to 0.04 in. The maximum angle of twist is limited to 0.3°.

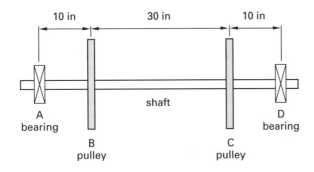

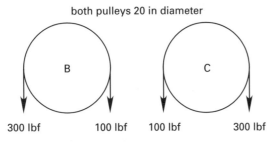

both pulleys 20 in diameter

REQUIREMENT

(a) Does the shaft satisfy the deflection limitation?

(b) Does the shaft satisfy the twist limitation?

(c) Draw the bending moment diagram and determine the maximum bending stress.

(d) What is the factor of safety for the shaft material assuming static loading?

(e) What is the factor of safety for the shaft material assuming completed reversed dynamic loading?

5 SITUATION

Two new concrete storage tanks are 100 ft apart and are being interconnected with 8 in nominal diameter schedule-40 steel pipe. All connections, joints, and fittings use flanged ends. During installation of the pipe, all flanged connections are securely made except the last connection, where a 0.5 in wide gap is noticed between the two flanges. Extra-long high-strength bolts are run through the flange holes, and the flanges are drawn together as the nuts are tightened. The ambient temperature at the time of installation is 80°F. The range of ambient temperatures seen by the pipe is 40°F to 110°F. Hot and cold water can be run through the pipe to perform proof testing at any desired temperature. The steel pipe has the following mechanical properties at all temperatures.

modulus of elasticity:	3×10^7 psi
yield strength:	30 ksi
Poisson's ratio:	0.3
coefficient of linear thermal expansion:	6.5×10^{-6} 1/°F

REQUIREMENT

(a) At what temperature should proof testing of the pipe be performed?

(b) What is the maximum longitudinal stress experienced by the pipe at 40°F?

(c) What are the principal stresses in the pipe if the pipe is tested at 40°F and 1000 psig?

(d) Disregarding longitudinal stress due to internal pressurization in the tank and pipe, will 1000 psig be a safe internal proof pressure if the proof temperature is 40°F?

(e) Why might the pipe have been intentionally constructed too short?

6 SITUATION

A typical clutch consisting of plates and disks is being designed to transmit 25 hp. The diameter of the clutch is established at 6 in, and each clutch disk and plate will have a 1 in diameter hole concentric with its center. Plates are steel and have a diameter of 0.2 in. When engaged, the shaft starts from zero and achieves 1500 rpm in 30 sec with uniform acceleration. The contact pressure when the clutch is fully engaged cannot exceed 100 psi. The coefficient of friction between the steel plate and disks is 0.1. The clutch material has an average density of 491 lbm/ft^3 and a specific heat of 0.107 Btu/lbm-°F. The material used for the clutch shaft has a tensile yield strength of 86 ksi. There is no significant bending moment on the clutch shaft. A factor of safety of 2.0 is to be used in all appropriate calculations.

REQUIREMENT

(a) What is the minimum shaft diameter?

(b) What is the minimum number of plates required? Is this a good design? Why or why not?

(c) What will be the average temperature of the clutch plates after one cycle of engagement? Is this temperature acceptable? State your assumptions.

(d) Would it be practical to use a typical clutch plate as a disk-brake rotor?

7 SITUATION

Five employees are currently needed to manufacture a particular metal product using a labor-intensive manual process. The employees are able to produce 12,000 units per month working 8-hour shifts for 5 days each week. Each employee is paid $15 per hour.

The manufacturer wants to replace all of the manual operations with a machine. Three different machines are available. One worker will still be needed to service the machines, however. That worker will be paid straight time (100%) at $15 per hour for the first 8 hours per day, and 150% of straight time for more than 8 hours up to 4 hours of overtime in any given week. If more than 4 hours of overtime are needed in a week, a swingshift employee will be brought in to work all hours in excess of 8 in a single shift instead of overtime being used. The swingshift employee will work a complete 8-hour shift at 125% of straight time.

Except for overtime, complete 8-hour shifts must be worked. Overtime is paid in increments of whole hours.

The three machines have the characteristics shown in the following table. Due to its customized nature, machine A will take 1 year to be built and delivered. Manual production will be required during that period. Machines B and C can be available in 6 months. Payment for each of the machines must be made at the time of the order for it. Because of their customized nature, none of the machines has a salvage value. Maintenance is paid semiannually in equal installments.

machine	initial cost	production rate	availability	yield	maintenance costs
A	$200,000	100/hr	90%	88%	$10,000/yr
B	$175,000	87.5/hr	80%	92%	$15,000/yr
C	$150,000	82/hr	75%	95%	$17,000/yr

The company uses an interest rate of 10% calculated semiannually. The company's business plan shows that the product will no longer be manufactured after 6 years. There are 8 hours in a normal workday, 5 days in a normal workweek, 52 weeks in a normal year, and 12 months in a normal year.

REQUIREMENT

(a) Determine the most economical alternative.

(b) Briefly describe how sensitive your decision is to the cost information.

8 SITUATION

A fire sprinkler system uses 1 in nominal diameter schedule-40 black steel pipe. Sprinklers are located 10 ft apart. The minimum pressure at any sprinkler is 10 psig. All sprinklers have standard $1/2$ in orifices with discharge coefficients of 0.75. The Hazen-Williams coefficient for the pipe is 120. In a particular event, the last three sprinklers on a branch line are fully open simultaneously.

All main lines and branches in the protected area are fed by a horizontal centrifugal pump. It is estimated that the demand for all fire streams (including sprinklers and hoses) will be 1300 gal/min at 90 psig. No significant suction head or lift is present.

REQUIREMENT

(a) What is the discharge from the most remote branch sprinkler?

(b) Disregarding velocity pressure, what is the discharge from the second sprinkler from the end?

(c) Including velocity pressure for all sprinklers, what is the discharge from the third sprinkler from the end?

(d) What minimum standard fire pump should be used?

9 SITUATION

A response of a second-order control system is described by the following differential equation.

$$\frac{d^2y}{dt^2} + 5\left(\frac{dy}{dt}\right) + 16y = 16x(t)$$

REQUIREMENT

(a) What is the characteristic equation in the s-domain?

(b) What is the undamped natural frequency?

(c) What is the damping ratio?

(d) What is the damped natural frequency?

(e) What is the time constant?

10 SITUATION

A sharp-edged orifice is placed in a water line. The orifice diameter is one-half of the inside pipe diameter. The temperature of the water is 60°F. The line is 12 in (nominal) diameter schedule-40 steel pipe. A static pressure gage located 20 diameters before the orifice plate shows that the pressure in the line is 200 psig. A differential mercury U-tube manometer placed across the orifice is used to determine the flow rate. The manometer is connected to the pipe through equal-length legs attached to smooth, flush wall taps.

REQUIREMENT

(a) In terms of pipe diameters before and after the orifice, where should the manometer taps be placed?

(b) If the orifice is to be placed immediately after a 90° bend in the line, how many pipe diameters past the bend should the orifice be located?

(c) What is the minimum Reynolds number such that the coefficient of flow for the orifice plate is constant?

(d) Given a flow rate of 1400 gal/min, what is the coefficient of flow?

(e) Given a flow rate of 1400 gal/min, what is the permanent pressure drop across the orifice plate?

(f) If the pressure difference across the wall taps is 6.0 lbf/in^2, what is the flow rate?

(g) The orifice plate is replaced with a venturi meter whose throat diameter is half of the pipe diameter. Given a pressure drop through the venturi at the throat of 6.0 lbf/in^2, what is the flow rate?

(h) Water is flowing in the pipe at 20 ft/sec when a valve is suddenly closed. What is the pressure increase in the pipe line?

(i) If the maximum pressure in the pipe reaches 500 lbf/in^2 due to the sudden valve closure, what is the maximum stress in the pipe?

STOP!

DO NOT CONTINUE!

This concludes PART 1 of the examination. PART 2 is the multiple-choice afternoon four-hour session. Once you begin PART 2, do not refer to PART 1 for any reason. If you finish PART 1 in less than four hours, it is suggested that you check your calculations.

Part 2—P.M.

11 SITUATION

A high-rise office building has a water tank located on the top floor. The tank is filled by a centrifugal pump located on the basement floor. Use the following information.

- The water temperature is 60°F.

- The kinematic viscosity of the water is 1.13 centistokes.

- The pump efficiency is 70%.

- The pump discharges 400 gal/min.

- The pressure of the feedwater entering the pump is 60 psig.

- The tank inlet is 806 ft above the pump outlet.

- The static pressure in the tank at its inlet is 60 psig.

- All pipe has an actual internal diameter of 6.065 in.

- The equivalent length of the pipe between the tank and pump is 910 ft.

- The relative roughness of the pipe is 0.001.

- The frictional head loss for all fittings and valves between the pump and tank is 1.15 ft of water.

REQUIREMENT

(a) What is the kinematic viscosity of the water?
- (A) 1.216×10^{-5} ft^2/sec
- (B) 1.216×10^{-3} ft^2/sec
- (C) 3.089×10^{-3} ft^2/sec
- (D) 0.175 ft^2/sec

(b) What is the velocity of the water through the pipe?
- (A) 2.3 ft/sec
- (B) 4.4 ft/sec
- (C) 14 ft/sec
- (D) 33 ft/sec

(c) What is the Reynolds number of the water in the pipe?
- (A) 8200
- (B) 12,400
- (C) 15,900
- (D) 18,500

(d) What is the pipe's friction factor?
- (A) 0.017
- (B) 0.019
- (C) 0.021
- (D) 0.025

(e) What is the head loss due to friction between the pump and the tank?
- (A) 1.2 ft
- (B) 3.9 ft
- (C) 11.7 ft
- (D) 12.7 ft

For questions (f) through (j), use the following data:

kinematic viscosity:	1.2×10^{-5} ft^2/sec
water bulk modulus:	320,000 psi
pipe inside diameter:	5.000 in
water velocity:	6.54 ft/sec
Reynolds number:	2.27×10^5
friction factor:	0.22
friction head loss:	33.1 ft
minor loss:	1.15 ft

(f) What is the total head against which the pump must work?
- (A) 32 ft
- (B) 810 ft
- (C) 840 ft
- (D) 910 ft

(g) What is the water's mass flow rate?
- (A) 1.7 lbm/sec
- (B) 1.7 poundal/sec
- (C) 55.6 lbm/sec
- (D) 55.6 slug/sec

(h) What is the theoretical power to the pump motor?

(A) 5 hp

(B) 85 hp

(C) 120 hp

(D) 130 hp

(i) Neglecting the elasticity of the pipe, what is the velocity of sound for water with the properties given?

(A) 34 ft/sec

(B) 410 ft/sec

(C) 860 ft/sec

(D) 4900 ft/sec

(j) What is the static pressure at the pump outlet just prior to turning on the pump if all valves are open and the supply tank is full?

(A) 60 psig

(B) 350 psig

(C) 410 psig

(D) 2600 psig

12 SITUATION

An ideal compressible fluid is flowing from point A to point B. The fluid's specific heat at constant volume is 0.17 Btu/lbm-°F, and its specific gas constant is 52.9 ft-lbf/lbm-°F. The properties of the fluid at points A and B are as follows:

	A	B
diameter	5 in	4.06 in
density	0.381 lbm/ft^3	0.0508 lbm/ft^3
velocity	190 ft/sec	2161 ft/sec
internal energy	119.1 Btu/lbm	53.1 Btu/lbm
pressure	98.4 psia	5.85 psia
elevation	0 ft	2 ft

REQUIREMENT

(a) What is the increase in kinetic energy between points A and B?

(A) 78 Btu/lbm

(B) 93 Btu/lbm

(C) 300 Btu/lbm

(D) 420 Btu/lbm

(b) What is the change in enthalpy between points A and B?

(A) −93 Btu/lbm (decrease)

(B) −66 Btu/lbm (decrease)

(C) −27 Btu/lbm (decrease)

(D) 77 Btu/lbm (increase)

(c) What is the change in flow energy (i.e., pV-work) between points A and B?

(A) −210 Btu/lbm (decrease)

(B) −27 Btu/lbm (decrease)

(C) 0.2 Btu/lbm (increase)

(D) 140 Btu/lbm (increase)

(d) What is the increase in potential energy between points A and B?

(A) 0.062 ft-lbf/lbm

(B) 0.10 ft-lbf/lbm

(C) 0.66 ft-lbf/lbm

(D) 2.0 ft-lbf/lbm

(e) What is the mass flow rate?

(A) 0.3 lbm/sec

(B) 10 lbm/sec

(C) 40 lbm/sec

(D) 320 lbm/sec

(f) If no work is done by the fluid on the surroundings, what is the total heat transfer (gain) between points A and B?

(A) 0.03 Btu/sec

(B) 2 Btu/sec

(C) 20 Btu/sec

(D) 30 Btu/sec

(g) If 20 Btu/lbm are added to the system between points A and B, what power is transferred to the surroundings?

(A) 180 ft-lbf/sec

(B) 280 ft-lbf/sec

(C) 1100 ft-lbf/sec

(D) 150,000 ft-lbf/sec

(h) What is the specific heat at constant pressure for the fluid?

(A) 0.068 Btu/lbm-°F

(B) 0.221 Btu/lbm-°F

(C) 0.238 Btu/lbm-°F

(D) 0.247 Btu/lbm-°F

(i) What would be the polytropic exponent if the fluid experiences a reversible adiabatic process?

(A) 1.20

(B) 1.29

(C) 1.33

(D) 1.40

(j) What kind of process is actually taking place between points A and B?

(A) isothermal heating

(B) throttling

(C) isothermal compression

(D) expansion through a nozzle

13 SITUATION

A portion of a steam-driven electrical generating power plant is shown. The turbine's isentropic efficiency is 87%. The efficiency of the boilerfeed pump is 100%. The furnace is 70% efficient and burns coal with 20% excess air by weight. The steam generator is 60% efficient.

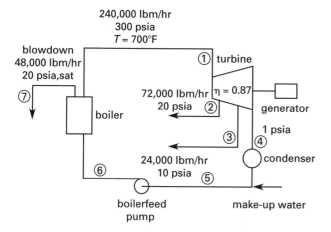

The coal is dry and has the following characteristics.

higher heating value:	13,500 Btu/lbm coal
% ash in fuel by weight:	10%
ash-free composition by weight:	
carbon	90%
hydrogen	4%
oxygen	4%
nitrogen	2%

Water leaves the condenser as a saturated 1 psia liquid. The blowdown is saturated steam at 20 psia. Atmospheric air is dry and is at 60°F and 14.7 psia. Combustion gases leave the stack at 60°F and 15 psia. Assume stack gases are ideal gases.

REQUIREMENT

(a) What is the enthalpy of the steam removed from the 20 psia (first) bleed?

(A) 1120 Btu/lbm

(B) 1150 Btu/lbm

(C) 1230 Btu/lbm

(D) 1240 Btu/lbm

(b) What is the maximum electrical power that can be generated?

(A) 9 kW

(B) 18 kW

(C) 22 kW

(D) 76 GW

(c) What is the steam flow rate through the condenser?

(A) 150 gal/min

(B) 200 gal/min

(C) 250 gal/min

(D) 300 gal/min

(d) What heat transfer occurs in the condenser?

(A) 100 MBtuh

(B) 130 MBtuh

(C) 170 MBtuh

(D) 220 MBtuh

(e) What work is done by the boilerfeed pump?

(A) 130 kBtuh

(B) 210 kBtuh

(C) 230 kBtuh

(D) 400 kBtuh

(f) What is the firing rate of the coal?

(A) 2800 lbm/hr

(B) 3700 lbm/hr

(C) 7.2×10^4 lbm/hr

(D) 6.5×10^5 lbm/hr

(g) What is the volumetric percentage of nitrogen in the stack gas?

(A) 36%

(B) 64%

(C) 77%

(D) 92%

(h) What is the maximum theoretical combustion temperature?

(A) 3200°F

(B) 3600°F

(C) 4000°F

(D) 4200°F

(i) What mass of water vapor is produced per pound of coal burned?

(A) 0.36

(B) 0.40

(C) 0.45

(D) 0.57

(j) If the stack pressure is 15.0 psia, what is the dew-point temperature of the water in the stack gas?

- (A) 85°F
- (B) 110°F
- (C) 135°F
- (D) 190°F

14 SITUATION

The compressor in an R-134a vapor compression refrigeration system has an isentropic efficiency of 80% and a mechanical efficiency of 90%. The refrigerant enters the compressor at 50 kPa with 20°C of superheat. At the compressor discharge, the pressure is 600 kPa. The condenser discharges 600 kPa saturated liquid to the evaporator coils through an expansion valve. The condenser is cooled by water that enters at 16°C and leaves at 22°C. The cooling capacity is 6 tons.

The compressor is of the single-acting reciprocating variety, with a single cylinder turning at 250 rpm. The stroke length is twice the cylinder bore diameter.

REQUIREMENT

(a) What is the enthalpy of the refrigerant entering the compressor?

- (A) 375 kJ/kg
- (B) 390 kJ/kg
- (C) 415 kJ/kg
- (D) 440 kJ/kg

(b) What is the enthalpy of the refrigerant entering the evaporator?

- (A) 390 kJ/kg
- (B) 410 kJ/kg
- (C) 430 kJ/kg
- (D) 460 kJ/kg

(c) What is the temperature of the refrigerant leaving the compressor?

- (A) −5°C
- (B) 14°C
- (C) 22°C
- (D) 295°C

(d) What is the refrigerant flow rate?

- (A) 0.13 kg/s
- (B) 0.18 kg/s
- (C) 0.27 kg/s
- (D) 0.45 kg/s

(e) Assuming that the refrigerant mass flow rate is 0.20 kg/s, what power is required to drive the compressor?

- (A) 5 kW
- (B) 11 kW
- (C) 16 kW
- (D) 20 kW

(f) Assuming that the refrigerant mass flow rate is 0.20 kg/s, what is the volumetric flow rate of the cooling water?

- (A) 0.1 L/s
- (B) 0.5 L/s
- (C) 2 L/s
- (D) 13 L/s

(g) What is the average specific heat of the refrigerant as it is being compressed?

- (A) 0.14 kJ/kg·K
- (B) 0.35 kJ/kg·K
- (C) 0.67 kJ/kg·K
- (D) 0.91 kJ/kg·K

(h) Assuming that the refrigerant mass flow rate is 0.20 kg/s, what is the volumetric flow rate of the refrigerant entering the compressor?

- (A) 80 L/s
- (B) 140 L/s
- (C) 170 L/s
- (D) 250 L/s

(i) Assuming that the refrigerant mass flow rate is 0.20 kg/s, what is the theoretical bore diameter of the cylinder?

- (A) 130 mm
- (B) 190 mm
- (C) 240 mm
- (D) 290 mm

(j) If the actual bore diameter was 310 mm, what would be the volumetric efficiency of the compressor?

- (A) 67%
- (B) 73%
- (C) 82%
- (D) 87%

15 SITUATION

By tracking down the cause of a process that has begun to operate at a degraded level, a plant engineer discovers a section of bare overhead steam pipe. Upon checking the plant's maintenance records, the engineer learns that a leaking steam trap had recently been repaired, and the saturated insulation had been removed from the pipe and discarded, but never replaced.

The properties of the pipe are as follows.

length of bare pipe section: 120 ft
pipe material: carbon steel
pipe size: 1½ in BWG 16 gage
pipe mounting: ceiling pipe hangers

Saturated steam at atmospheric pressure flows through the pipe at a high enough rate to prevent substantial condensation. The inside heat transfer coefficient is 1500 Btu/hr-ft²-°F. The outside heat transfer coefficient for the bare pipe in still air is 2.0 Btu/hr-ft²-°F.

The air in the plant is at 60°F and 14.7 psia and is normally still. The pipe temperature is too low to consider the effects of radiation.

REQUIREMENT

(a) What is the approximate thermal conductivity of the pipe?

- (A)   20 Btu/hr-ft-°F
- (B)   25 Btu/hr-ft-°F
- (C)   30 Btu/hr-ft-°F
- (D)   120 Btu/hr-ft-°F

(b) Assuming a thermal conductivity of 29 Btu/hr-ft-°F, what is the thermal resistance of a 1 ft long section of pipe?

- (A)   0.00005 hr-°F/Btu
- (B)   0.0005 hr-°F/Btu
- (C)   0.0017 hr-°F/Btu
- (D)   0.039 hr-°F/Btu

(c) What is the overall heat transfer coefficient (based on the outside diameter) in still air?

- (A)   0.5 Btu/hr-ft²-°F
- (B)   2 Btu/hr-ft²-°F
- (C)   4 Btu/hr-ft²-°F
- (D)   270 Btu/hr-ft²-°F

(d) How much heat is lost from the bare pipe in 45 min per volume of pipe?

- (A)   8800 Btu/ft³
- (B)   10,800 Btu/ft³
- (C)   11,600 Btu/ft³
- (D)   12,800 Btu/ft³

(e) In a particular procedure, valves at each end of the pipe section are closed instantaneously, and the steam in the pipe begins to cool. A vacuum breaker prevents any substantial vacuum from forming in the pipe. The average heat transfer over the cooling period is 27% of the maximum bare-pipe rate with the valves open. There is no heat loss through the valves at the ends of the pipe, or through the vacuum breaker. How long does it take for the steam to cool to the ambient temperature of 60°F?

- (A)   1 min
- (B)   12 min
- (C)   48 min
- (D)   110 min

(f) What is the effective thermal resistance per foot of pipe length of a 2 in thick layer of 85% magnesia insulation wrapped around the pipe?

- (A)   0.0059 °F-sec/Btu
- (B)   0.19 °F-sec/Btu
- (C)   0.26 °F-sec/Btu
- (D)   6.5 °F-sec/Btu

(g) When a 2 in thick layer of 85% magnesia insulation is wrapped around the pipe, the outside temperature of the insulation is determined to be 70°F. What is the rate of heat transfer from the insulated pipe?

- (A)   850 Btuh
- (B)   1100 Btuh
- (C)   2400 Btuh
- (D)   4500 Btuh

(h) If a ceiling fan is installed above the bare pipe, the maximum air velocity over the pipe can reach 30 ft/sec. Which of the following expressions would represent the Nusselt number for forced convection over the pipe?

- I. $C(\text{Re})^\alpha$
- II. $C(\text{Re})^\alpha (\text{Pr})^\beta$
- III. $0.023(\text{Re})^{0.8}(\text{Pr})^{0.3}$

- (A)   I only
- (B)   II only
- (C)   I and II
- (D)   II and III

(i) If the ceiling fan is operating as described in part (h), what is the approximate outside film coefficient?

- (A)   4.2 Btu/hr-ft²-°F
- (B)   10 Btu/hr-ft²-°F
- (C)   32 Btu/hr-ft²-°F
- (D)   60 Btu/hr-ft²-°F

(j) Approximately how much would the value of the outside film coefficient increase if the air velocity was doubled?

- (A) 10%
- (B) 50%
- (C) 100%
- (D) 400%

16 SITUATION

15,000 ft^3/min of air at 80°F dry bulb and 70% relative humidity enter an air conditioner and are discharged at 56°F dry bulb and 90% relative humidity. 3000 ft^3/min of outside air at 90°F and 45% relative humidity bypass the air conditioner and are mixed with the air leaving the air conditioner. The mixture enters the conditioned space.

REQUIREMENT

(a) What is the wet-bulb temperature of the outside air?

- (A) 65°F
- (B) 69°F
- (C) 73°F
- (D) 77°F

(b) What is the total mass of moisture brought in by the outside air?

- (A) 2.8 lbm/hr
- (B) 15 lbm/hr
- (C) 85 lbm/hr
- (D) 170 lbm/hr

(c) What is the dew point of the outside air?

- (A) 52°F
- (B) 57°F
- (C) 65°F
- (D) 77°F

(d) What is the sensible heat factor for the air conditioning process?

- (A) 0.29
- (B) 0.36
- (C) 0.45
- (D) 0.66

(e) What mass of water is removed by the air conditioner?

- (A) 7.3 lbm/hr
- (B) 90 lbm/hr
- (C) 440 lbm/hr
- (D) 610 lbm/hr

(f) What is the dew point of the air mixture?

- (A) 54°F
- (B) 55°F
- (C) 58°F
- (D) 62°F

(g) What is the specific humidity of the air mixture?

- (A) 0.0012 lbm/lbm
- (B) 0.0014 lbm/lbm
- (C) 0.0016 lbm/lbm
- (D) 0.0094 lbm/lbm

(h) What is the dry-bulb temperature of the air mixture?

- (A) 62°F
- (B) 65°F
- (C) 69°F
- (D) 74°F

(i) If the air mixture enters the conditioned space where the sensible heat load is 2×10^5 Btu/hr, what will be the exit temperature of the air?

- (A) 70°F
- (B) 72°F
- (C) 75°F
- (D) 77°F

(j) If the air mixture leaving the conditioned space is at 75°F dry bulb and 70% relative humidity, what fraction of the room heating load is latent heating?

- (A) 0.35
- (B) 0.45
- (C) 0.55
- (D) 0.65

17 SITUATION

While planning for the renovation of an old residential condominium complex, an opportunity has arisen to replace old double-hung windows with newer casement windows of the same size. The expected benefits are a less obstructed view and a reduction in infiltration. There are 10 double-hung windows per condominium.

The following information is available.

location:	New York City (near La Guardia airport)
summer outside design temperature (2.5% exceedance):	89°F, dry bulb
summer inside design temperature:	80°F, dry bulb 50% relative humidity
summer design wind speed:	15 mph (pressure of 0.10 in wg)
length of summer cooling season:	3 months
winter outside design temperature (97.5% exceedance):	15°F, dry bulb
winter inside design temperature:	70°F, dry bulb
condominiums in the complex:	20
exterior wall construction:	masonry block; caulked window frames
double-hung window:	3 ft × 5 ft; loose fit; weatherstripped
casement window:	wood frame; ANSI type A200.1; class A; 3 ft × 5 ft; crack length: 16 ft
summer cooling equipment:	unitary through-wall units
summer energy efficiency ratio (EER):	8 Btu/W-hr
cost of electricity:	$0.15/kW-hr; $20/kW-month demand charge
air-conditioning system— summer equivalent full-load hours:	1000 hr
winter heating fuel:	#6 fuel oil
cost of fuel oil:	$1.50/gal
furnace efficiency:	55%

REQUIREMENT

(a) What is the crack length for each casement window?

(A) 3 ft

(B) 16 ft

(C) 19 ft

(D) 20 ft

(b) The infiltration rate per casement window is closest to

(A) 200 ft^3/hr

(B) 300 ft^3/hr

(C) 400 ft^3/hr

(D) 500 ft^3/hr

(c) Assuming that the infiltration for a casement window is 430 ft^3/hr, what is the difference in infiltration per window achieved by replacing the double-hung windows with casement windows?

(A) 80 ft^3/hr

(B) 120 ft^3/hr

(C) 150 ft^3/hr

(D) 180 ft^3/hr

(d) Assuming that the difference in infiltration per window is 120 ft^3/hr, what reduction in summer peak heat gain per window is achieved by replacing the double-hung windows with casement windows?

(A) 20 Btu/hr

(B) 40 Btu/hr

(C) 70 Btu/hr

(D) 100 Btu/hr

(e) What is the inside design humidity ratio?

(A) 0.008 lbm/lbm

(B) 0.011 lbm/lbm

(C) 0.014 lbm/lbm

(D) 0.018 lbm/lbm

(f) Assuming that the outside design conditions are 89°F dry bulb and 75°F wet bulb, and that the difference in infiltration per window is 120 ft^3/hr, what is the reduction in peak summer latent heat gain due to replacing the double-hung windows with casement windows?

(A) 20 Btu/hr

(B) 40 Btu/hr

(C) 60 Btu/hr

(D) 80 Btu/hr

(g) Assuming that the total (sensible and latent) peak summer heat gain is 40 Btu/hr per window, what is the annual value of the energy savings per condominium due to replacing the double-hung windows with casement windows?

(A) $8 per year

(B) $14 per year

(C) $27 per year

(D) $49 per year

(h) Assuming that the total (sensible and latent) peak summer heat gain is 40 Btu/hr per window, what is the annual reduction in demand charges per condominium due to replacing the double-hung windows with casement windows?

(A) $3 per year

(B) $9 per year

(C) $18 per year

(D) $60 per year

(i) Assuming that the difference in infiltration per window is 120 ft^3/hr, what is the reduction in winter infiltration sensible heat loss per window?

(A) 90 Btu/hr

(B) 120 Btu/hr

(C) 140 Btu/hr

(D) 220 Btu/hr

(j) Assuming that the reduction in winter heat loss is 200 Btu/hr per window, what are the approximate winter energy savings per condominium due to replacing the double-hung windows with casement windows?

(A) $7 per year

(B) $20 per year

(C) $50 per year

(D) $80 per year

18 SITUATION

A sheet-metal flat spring is constructed as shown. The force is applied in the x-direction at point D. The material is AISI 430 stainless steel, cold rolled, and with an annealed temper and the following properties.

thickness:	0.040 in
ultimate strength:	75 ksi
yield strength:	55 ksi
endurance limit	
(rotating beam specimen):	30 ksi
modulus of elasticity:	29×10^6 psi
Poisson's ratio:	0.30
theoretical tensile stress	
concentration factor from	
bend radius at point B (K_t):	1.4
notch sensitivity factor (q):	0.80

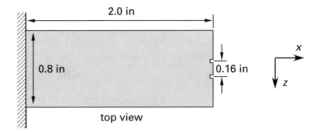

top view

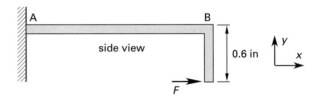

side view

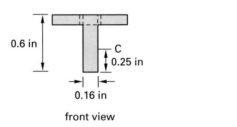

front view

REQUIREMENT

(a) If the applied force is a constant 1.5 lbf, the maximum normal stress at point A is

(A) −4170 psi (compression)

(B) 4220 psi (tension)

(C) 4270 psi (tension)

(D) 5910 psi (tension)

(b) If the applied force is a constant 2 lbf, the deflection of point D in the x-direction with respect to point B is

(A) 0.0012 in

(B) 0.0047 in

(C) 0.0053 in

(D) 0.0058 in

(c) If the applied force is a constant 1.75 lbf, the deflection of point B in the y-direction with respect to point A is closest to

(A) 0.005 in

(B) 0.009 in

(C) 0.015 in

(D) 0.017 in

(d) If the applied force is a constant 2 lbf, the principal stress at point B is

(A) 5600 psi

(B) 16,000 psi

(C) 28,000 psi

(D) 39,000 psi

(e) If the applied force is a constant 2 lbf, the maximum shear stress at point C is

(A) 310 psi

(B) 470 psi

(C) 1200 psi

(D) 5900 psi

(f) If the applied force alternates between 0 and 1 lbf, and if the theoretical stress concentration factor is disregarded, the alternating principal stress at point B is

(A) 5700 psi

(B) 7000 psi

(C) 8100 psi

(D) 9800 psi

(g) If the applied force alternates between 0 and 1 lbf, and if the theoretical stress concentration factor is disregarded, the mean principal stress at point B is

(A) 5100 psi

(B) 5700 psi

(C) 7000 psi

(D) 8100 psi

(h) At one particular loading, the mean stress is 35,000 psi and the alternating stress is 12,000 psi at point B, as predicted by the distortion energy theory. The factor of safety based on a modified Goodman diagram (modified Goodman line) is

(A) 0.93

(B) 1.0

(C) 1.9

(D) 2.1

(i) The manufacturing process leaves a residual tensile stress at point B of 32 ksi. For a certain cycling loading, the minimum and maximum tensile stresses are 3515 psi and 21,094 psi, respectively. The resulting alternating and mean stresses are, respectively,

(A) −8800 psi (compression);
     −20,000 psi (compression)

(B) 8800 psi (tension); 44,000 psi (tension)

(C) 25,000 psi (tension); 22,000 psi (tension)

(D) 36,000 psi (tension); 53,000 psi (tension

(j) It is known that the endurance limit for this material comes from a normally distributed population with a mean of 30 ksi and a variance of 5760×10^3 lbf^2/in^4. What is the probability that a spring's endurance limit will be at least 23,250 psi?

(A) 87.9%

(B) 94.0%

(C) 98.9%

(D) 99.8%

19 SITUATION

A 3 in diameter, solid circular shaft is simply supported by two ball bearings at A and D, as shown. The shaft carries two pulleys at B and C. Pulley B has a diameter of 15 in and carries the vertical loads shown. Pulley C has a diameter of 12 in and carries the horizontal loads shown. Point E is located 10 in from bearing A. The shaft and pulley masses are insignificant, as are the centrifugal effects of the belt. (Refer to the illustration for the direction of the axes.)

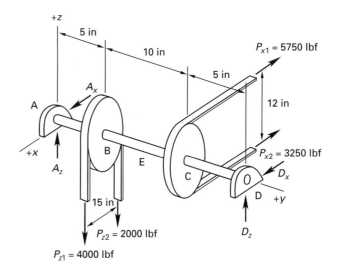

REQUIREMENT

(a) If the contact angle for pulley C is 180°, what is the coefficient of friction between the belt and the pulley?

(A) 0.08

(B) 0.18

(C) 0.22

(D) 0.57

(b) What is the reaction at bearing A in the x-direction?

(A) −9000 lbf

(B) 0 lbf

(C) 2250 lbf

(D) 6750 lbf

(c) What is the magnitude of the bending moment around the x-axis acting on point E?

(A) 15,000 in-lbf

(B) 23,000 in-lbf

(C) 30,000 in-lbf

(D) 38,000 in-lbf

(d) What is the torque in the shaft at point E?

(A) 0 in-lbf

(B) 15,000 in-lbf

(C) 30,000 in-lbf

(D) 45,000 in-lbf

(e) What is the magnitude of the direct shear in the z-direction at point E?

(A) 0 lbf

(B) 1500 lbf

(C) 2500 lbf

(D) 6000 lbf

For questions (f) through (j), use the following information for the moments and shear acting on point E. (Subscripts refer to the axis about which rotation is directed. Refer to the illustration for the direction of the axes.)

$$M_x = 22{,}500 \text{ in-lbf}$$
$$M_y = 25{,}000 \text{ in-lbf}$$
$$M_z = 20{,}000 \text{ in-lbf}$$
$$V_z = 5000 \text{ lbf}$$

(f) What is the maximum bending stress for a point on the shaft located at $(x, y, z) = (0, 10 \text{ in}, 1.5 \text{ in})$?

(A) 0 psi

(B) 940 psi

(C) 1400 psi

(D) 8500 psi

(g) What is the maximum shear stress in the z-direction at point E?

(A) 0 psi

(B) 350 psi

(C) 710 psi

(D) 940 psi

(h) What is the maximum torsional shear stress at point E?

(A) 590 psi

(B) 880 psi

(C) 3500 psi

(D) 4700 psi

(i) What is the maximum shear stress in the shaft at a point located at $(x, y, z) = (1.5 \text{ in}, 0, 0)$?

(A) 99 psi

(B) 4700 psi

(C) 5400 psi

(D) 6000 psi

(j) What is the maximum normal stress in the shaft at a point located at $(x, y, z) = (1.5 \text{ in}, 0, 0)$?

(A) 940 psi

(B) 3000 psi

(C) 5700 psi

(D) 9800 psi

20 SITUATION

An anniversary clock contains a rotary pendulum. The pendulum can be modeled as three steel balls spinning around a vertical axis, acted upon by an ideal torsional spring. The horizontal arms supporting the balls are rigid and massless. All friction and other external forces acting on the pendulum to retard its motion are counteracted by the clock's battery. The following information is available.

   I.   torsional spring
           constant: 2.857×10^{-3} in-lbf/rad
  II.   steel density: 0.284 lbm/in^3
 III. ball diameter: 0.400 in
 IV. steel modulus of
           elasticity: 2.9×10^6 psi
  V.   dimension A: 2.3 in
 VI. dimension B: 0.75 in

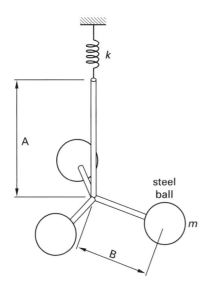

steel
ball

m

REQUIREMENT

(a) What information is required to determine the natural frequency of the system?

(A) I, III, IV, and V

(B) I, II, III, and V

(C) I, II, III, and VI

(D) I, V, and VI

(b) If the pendulum is initially wound until the torque in the spring is 0.090 in-lbf and then released, what will be the amplitude of oscillation?

(A) 0.8 rev

(B) 1 rev

(C) 2 rev

(D) 5 rev

(c) As described, the motion of the pendulum is an example of

(A) damped free vibration

(B) undamped free vibration

(C) critically damped free vibration

(D) undamped forced vibration

(d) If a different torsional spring was used such that the spring constant was doubled, what would be the effect?

(A) The natural frequency would be reduced by about 50%.

(B) The natural frequency would be doubled.

(C) The natural frequency would be increased by about 40%.

(D) The amplitude of oscillation would be decreased by about 15%.

(e) If the clock was transported to a location where the acceleration due to gravity was 8 ft/sec^2, what would be the effect?

(A) The natural frequency would be reduced by about 50%.

(B) The natural frequency would be reduced by about 24%.

(C) The natural frequency would be increased by about 50%.

(D) The natural frequency would be unchanged.

(f) When the amplitude of oscillation is 720°, the maximum torque on the spring will be most nearly

(A) 0.015 in-lbf

(B) 0.036 in-lbf

(C) 0.080 in-lbf

(D) 440 in-lbf

(g) Which of the following is not a consequence of increasing dimension B?

(A) The natural frequency will increase.

(B) The period of oscillations will decrease.

(C) The spring constant must be decreased to maintain the same natural frequency.

(D) The mass of the balls must be reduced to maintain the same natural frequency.

(h) As soon as the clock's battery is disconnected, which of the following will occur?

(A) The natural frequency will slowly decrease to zero, while the amplitude of oscillation remains the same.

(B) The amplitude of oscillation will slowly decrease to zero while the natural frequency remains the same.

(C) The natural frequency and the amplitude of oscillation will both slowly decrease to zero.

(D) The natural frequency and the amplitude of oscillation will both remain the same.

(i) As described in part (h), the motion of the pendulum is an example of

(A) undamped free vibration

(B) damped free vibration

(C) damped resonant vibration

(D) damped forced vibration

(j) The frequency of oscillation is most nearly

(A) 1.3 Hz

(B) 2.1 Hz

(C) 2.7 Hz

(D) 8.2 Hz

Solutions Part 1

1 SOLUTION

(a) Evaluate the properties of the air at point 2, the entrance to the turbine.

$$p_2 = 100 \text{ lbf/in}^2 \quad \text{[absolute]}$$
$$T_2 = 250°\text{F} + 460 = 710°\text{R}$$

From a low-pressure air property table for 710°R,

$$h_2 = 169.98 \text{ Btu/lbm}$$

The volumetric flow rate is

$$\dot{V}_2 = \frac{\dot{m}RT}{p}$$

$$= \frac{\left(68 \dfrac{\text{lbm}}{\text{min}}\right)\left(53.3 \dfrac{\text{ft-lbf}}{\text{lbm-°R}}\right)(710°\text{R})}{\left(100 \dfrac{\text{lbf}}{\text{in}^2}\right)\left(144 \dfrac{\text{in}^2}{\text{ft}^2}\right)}$$

$$= 178.7 \text{ ft}^3/\text{min}$$

Now, evaluate the properties of the air at point 3, the exit of the turbine. The expansion is polytropic.

$$p_2 V_2^n = p_3 V_3^n$$

$$\dot{V}_3 = \dot{V}_2 \left(\frac{p_2}{p_3}\right)^{\frac{1}{n}} = \left(178.7 \; \frac{\text{ft}^3}{\text{min}}\right)\left(\frac{100 \; \frac{\text{lbf}}{\text{in}^2}}{12 \; \frac{\text{lbf}}{\text{in}^2}}\right)^{\frac{1}{1.2}}$$

$$= 1045.9 \text{ ft}^3/\text{min}$$

The ideal temperature is

$$T_3 = T_2 \left(\frac{p_3}{p_2}\right)^{\frac{n-1}{n}} = (710°\text{R})\left(\frac{12 \; \frac{\text{lbf}}{\text{in}^2}}{100 \; \frac{\text{lbf}}{\text{in}^2}}\right)^{\frac{1.2-1}{1.2}}$$

$$= 498.6°\text{R} \quad (38.6°\text{F})$$

From a low-pressure air property table, the enthalpy is

$$h_3 = 119.1 \text{ Btu/lbm}$$

The actual turbine work is

$$W_{2-3} = \eta W_{\text{ideal}} = \eta \dot{m}(h_2 - h_3)$$

$$= \frac{(0.85)\left(68 \; \dfrac{\text{lbm}}{\text{min}}\right)}{\left(1000 \; \dfrac{\text{W}}{\text{kW}}\right)\left(3.413 \; \dfrac{\text{Btu}}{\text{W-hr}}\right)}$$
$$\times \left(169.98 \; \dfrac{\text{Btu}}{\text{lbm}} - 119.1 \; \dfrac{\text{Btu}}{\text{lbm}}\right)\left(60 \; \dfrac{\text{min}}{\text{hr}}\right)$$

$$= 51.7 \text{ kW}$$

(b) When the air leaves the cabin, its properties are

$$p_4 = 12 \text{ lbf/in}^2$$
$$T_4 = 80°\text{F} + 460 = 540°\text{R}$$
$$h_4 = 129.06 \text{ Btu/lbm}$$

The cooling effect in tons of cooling is

$$Q = \dot{m}(h_4 - h_3)$$

$$= \frac{\left(68 \; \dfrac{\text{lbm}}{\text{min}}\right)\left(129.06 \; \dfrac{\text{Btu}}{\text{lbm}} - 119.1 \; \dfrac{\text{Btu}}{\text{lbm}}\right)}{200 \; \dfrac{\text{Btu}}{\text{min-ton}}}$$

$$= 3.386 \text{ tons}$$

(c) If the air was an ideal gas, its specific heat would be constant. The expansion can be isentropic even though the mechanical efficiency is less than 100%. For an isentropic expansion,

$$W_{2-3} = \eta \dot{m} c_p T_2 \left(1 - \frac{p_3}{p_2}\right)^{\frac{k-1}{k}}$$

$$= \left[\frac{(0.85)\left(68 \; \dfrac{\text{lbm}}{\text{min}}\right)\left(0.24 \; \dfrac{\text{Btu}}{\text{lbm-°R}}\right)(710°\text{R})}{3413 \; \dfrac{\text{Btu}}{\text{kW-hr}}}\right]$$

$$\times \left[1 - \left(\frac{12 \; \frac{\text{lbf}}{\text{in}^2}}{100 \; \frac{\text{lbf}}{\text{in}^2}}\right)^{\frac{1.4-1}{1.4}}\right]\left(60 \; \frac{\text{min}}{\text{hr}}\right)$$

$$= 78.67 \text{ kW}$$

(d) For an isentropic process,

$$T_3 = T_2 \left(\frac{p_3}{p_2}\right)^{\frac{k-1}{k}}$$

$$= (710°\text{R}) \left(\frac{12 \frac{\text{lbf}}{\text{in}^2}}{100 \frac{\text{lbf}}{\text{in}^2}}\right)^{\frac{1.4-1}{1.4}}$$

$$= 387.4°\text{R} \quad (-72.6°\text{F})$$

The ideal isentropic cooling effect is

$$Q = \dot{m}c_p(T_4 - T_3)$$

$$= \frac{\left(68 \frac{\text{lbm}}{\text{min}}\right)\left(0.24 \frac{\text{Btu}}{\text{lbm-°R}}\right)(540°\text{R} - 387.4°\text{R})}{200 \frac{\text{Btu}}{\text{min-ton}}}$$

$$= 12.45 \text{ tons}$$

However, air could not be introduced into the passenger cabin at such a low temperature.

2 SOLUTION

(a) First, determine the properties of the original configuration.

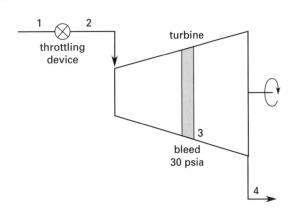

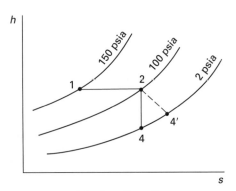

original configuration

From steam tables or a Mollier diagram, the enthalpy of the incoming 150°F steam is

$$h_1 = 1274 \text{ Btu/lbm}$$

Use a Mollier diagram to determine the properties of the throttled steam. In an ideal throttling process, enthalpy is constant. Constant-enthalpy processes are horizontal lines on the Mollier diagram. Therefore, move horizontally to the right on the Mollier diagram until the 100 psia line is reached.

$$h_2 = 1274 \text{ Btu/lbm}$$

In an isentropic (ideal) expansion process, entropy is constant. Constant-entropy processes are vertical lines on the Mollier diagram. Drop straight down to the 2 psia line.

$$h_4 = 989 \text{ Btu/lbm}$$

The actual turbine work is

$$W_{\text{actual},2-4} = \eta W_{\text{ideal},2-4} = \eta(h_2 - h_4)$$

$$= (0.86)\left(1274 \frac{\text{Btu}}{\text{lbm}} - 989 \frac{\text{Btu}}{\text{lbm}}\right)$$

$$= 245 \text{ Btu/lbm}$$

Now, evaluate the properties in the new configuration.

$$h_1 = 1274 \text{ Btu/lbm} \quad \text{[unchanged]}$$

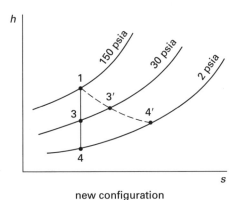

new configuration

From the original point, drop straight down to the 30 psia line to the ideal point 3.

$$h_3 = 1136 \text{ Btu/lbm} \quad \text{[from Mollier diagram at 30 psia]}$$

The actual enthalpy at point 3 is

$$h_3' = h_1 - \eta(h_1 - h_3)$$

$$= 1274 \frac{\text{Btu}}{\text{lbm}} - (0.86)\left(1274 \frac{\text{Btu}}{\text{lbm}} - 1136 \frac{\text{Btu}}{\text{lbm}}\right)$$

$$= 1155 \text{ Btu/lbm}$$

Continue straight down to the 2 psia line to the ideal point 4.

$$h_4 = 964 \text{ Btu/lbm} \quad \text{[from Mollier diagram at 2 psia]}$$

The actual enthalpy at point 4 is

$$h_4' = h_1 - \eta(h_1 - h_4)$$
$$= 1274\,\frac{\text{Btu}}{\text{lbm}} - (0.86)\left(1274\,\frac{\text{Btu}}{\text{lbm}} - 964\,\frac{\text{Btu}}{\text{lbm}}\right)$$
$$= 1007 \text{ Btu/lbm}$$

With a fraction, x, of the steam being bled off, the actual turbine work is

$$W' = h_1 - h_3' + (1-x)(h_3 - h_4')$$

Solve this for the unknown bleed fraction.

$$245\,\frac{\text{Btu}}{\text{lbm}} = 1274\,\frac{\text{Btu}}{\text{lbm}} - 1155\,\frac{\text{Btu}}{\text{lbm}}$$
$$+ (1-x)\left(1155\,\frac{\text{Btu}}{\text{lbm}} - 1007\,\frac{\text{Btu}}{\text{lbm}}\right)$$
$$x = 0.149 \quad (14.9\%)$$

(b) The enthalpy of liquid water does not depend significantly on pressure. The enthalpy of saturated 60°F water is found from the steam tables as

$$h_{\text{water}} = 28.06 \text{ Btu/lbm}$$

The steam that is not bled off passes through the condenser and constitutes the water passing through the feedwater heater that is heated by the bleed.

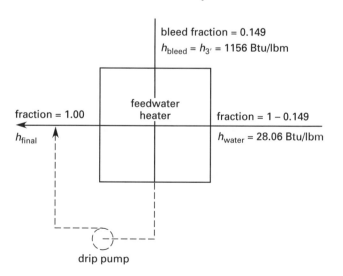

bleed fraction = 0.149
$h_{\text{bleed}} = h_{3'} = 1156$ Btu/lbm

feedwater heater

fraction = 1.00

h_{final}

fraction = 1 − 0.149

$h_{\text{water}} = 28.06$ Btu/lbm

drip pump

Neglecting drip pump work, the final mixture enthalpy is

$$h_{\text{final}} = xh_3' + (1-x)h_{\text{water}}$$
$$= (0.149)\left(1155\,\frac{\text{Btu}}{\text{lbm}}\right) + (1-0.149)\left(28.06\,\frac{\text{Btu}}{\text{lbm}}\right)$$
$$= 196 \text{ Btu/lbm}$$

Locating this enthalpy in the saturated liquid column of the steam tables, the temperature is approximately

$$T_{\text{final}} = 228°\text{F}$$

3 SOLUTION

(a) Refer to the figure. The broken lines are not relevant to the solution.

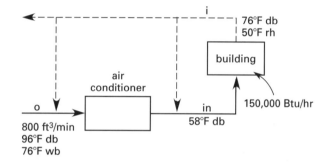

i
76°F db
50°F rh

building

air conditioner

o
800 ft³/min
96°F db
76°F wb

in
58°F db

150,000 Btu/hr

The required airflow rate through the room can be found from the sensible heat rise.

$$\dot{V}_{\text{cfm}} = \frac{q_s}{c_p \rho \Delta T}$$
$$= \frac{q_{s,\text{Btu/hr}}}{\left(1.08\,\frac{\text{Btu-min}}{\text{ft}^3\text{-hr-}°\text{F}}\right)(T_i - T_{\text{in}})}$$
$$= \frac{(0.8)\left(150{,}000\,\dfrac{\text{Btu}}{\text{hr}}\right)}{\left(1.08\,\dfrac{\text{Btu-min}}{\text{ft}^3\text{-hr-}°\text{F}}\right)(76°\text{F} - 58°\text{F})}$$
$$= 6173 \text{ ft}^3/\text{min}$$

(b) Use the psychrometric chart. Locate point "i" at the intersection of 76°F dry bulb and 50% relative humidity. The sensible heat ratio (i.e., sensible heat factor) is 0.8. Draw a line through point "i" with the psychrometric slope of 0.8. This is the process line describing the condition of the air as it moves through the room. The incoming air is known to have a dry-bulb temperature of 58°F, so locate the intersection of the process line and the vertical 58°F dry-bulb line. This is point "in". Read the specific humidity.

$$\omega = 60 \text{ gr/lbm} \quad (0.00857 \text{ lbm/lbm})$$

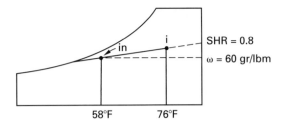

(c) Use the psychrometric chart to determine the properties of the air at points "o" and "i".

$$h_o = 39.4 \text{ Btu/lbm}$$
$$v_o = 13.45 \text{ ft}^3/\text{lbm}$$
$$h_i = 28.7 \text{ Btu/lbm}$$

The total air conditioning load includes the load from the outside air.

$$q_c = q_t + \dot{m}(h_o - h_i) = q_t + \left(\frac{\dot{V}_o}{v_o}\right)(h_o - h_i)$$

$$= 150,000 \frac{\text{Btu}}{\text{lbm}} + \left[\frac{\left(800 \frac{\text{ft}^3}{\text{min}}\right)\left(60 \frac{\text{min}}{\text{hr}}\right)}{13.45 \frac{\text{ft}^3}{\text{lbm}}}\right]$$

$$\times \left(39.4 \frac{\text{Btu}}{\text{lbm}} - 28.7 \frac{\text{Btu}}{\text{lbm}}\right)$$

$$= 1.882 \times 10^5 \text{ Btu/hr}$$

Convert this to tons.

$$q_c = \frac{1.882 \times 10^5 \frac{\text{Btu}}{\text{hr}}}{12,000 \frac{\text{Btu}}{\text{hr-ton}}}$$

$$= 15.68 \text{ tons}$$

4 SOLUTION

(a) Each pulley supports a downward force of

$$F = 100 \text{ lbf} + 300 \text{ lbf} = 400 \text{ lbf}$$

The area moment of inertia for a round shaft is

$$I = \tfrac{1}{4}\pi r^4 = \tfrac{1}{4}\pi(1 \text{ in})^4$$
$$= 0.7854 \text{ in}^4$$

Since the steel is hardened, use a modulus of elasticity of 3×10^7 lbf/in^2.

For a beam of length L with two equal loads of F each located a distance a from the ends, the maximum deflection occurs at the midpoint and is

$$y_{\max} = \left(\frac{Fa}{24EI}\right)(3L^2 - 4a^2)$$

$$= \left[\frac{(400 \text{ lbf})(10 \text{ in})}{(24)\left(3 \times 10^7 \frac{\text{lbf}}{\text{in}^2}\right)(0.7854 \text{ in}^4)}\right]$$

$$\times \left[(3)(50 \text{ in})^2 - (4)(10 \text{ in})^2\right]$$

$$= 0.0502 \text{ in}$$

Since $y_{\max} > 0.04$ in, the shaft does not meet the deflection specifications.

(b) The torque on each pulley is

$$T = (F_1 - F_2)r = (300 \text{ lbf} - 100 \text{ lbf})\left(\frac{20 \text{ in}}{2}\right)$$

$$= 2000 \text{ in-lbf}$$

The shear modulus is

$$G = \frac{E}{2(1+\mu)} = \frac{3 \times 10^7 \frac{\text{lbf}}{\text{in}^2}}{(2)(1+0.283)}$$

$$= 1.17 \times 10^7 \text{ lbf/in}^2$$

The polar area moment of inertia is

$$J = \tfrac{1}{2}\pi r^4 = \tfrac{1}{2}\pi\left(\frac{2 \text{ in}}{2}\right)^4$$

$$= 1.571 \text{ in}^4$$

There is no torque in the shaft outboard of the two pulleys. The twist in the shaft between the two pulleys is

$$\phi = \frac{TL}{GJ}$$

$$= \frac{(2000 \text{ in-lbf})(30 \text{ in})}{\left(1.17 \times 10^7 \frac{\text{lbf}}{\text{in}^2}\right)(1.571 \text{ in}^4)}$$

$$= 3.26 \times 10^{-3} \text{ rad}$$

$$\theta = \phi\left(\frac{360°}{2\pi}\right)$$

$$= (3.26 \times 10^{-3} \text{ rad})\left(\frac{360°}{2\pi}\right)$$

$$= 0.187°$$

Since $\theta < 0.3°$, the shaft does meet the twist limitation.

(c) Since the vertical loading is symmetrical, each bearing reaction is 400 lbf. The moment at one pulley due to the vertical forces on the other pulley and the opposite reaction is

$$M = Fx = (400 \text{ lbf})(40 \text{ in}) - (400 \text{ lbf})(30 \text{ in})$$
$$= 4000 \text{ in-lbf}$$

The bending moment diagram is

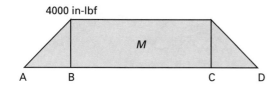

4000 in-lbf

The maximum bending stress is

$$\sigma = \frac{Mc}{I} = \frac{(4000 \text{ in-lbf})\left(\dfrac{2 \text{ in}}{2}\right)}{0.7854 \text{ in}^4}$$
$$= 5093 \text{ lbf/in}^2$$

(d) The maximum torsional shear stress in the shaft is

$$\tau = \frac{Tr}{J} = \frac{(2000 \text{ in-lbf})\left(\dfrac{2 \text{ in}}{2}\right)}{1.571 \text{ in}^4}$$
$$= 1273 \text{ lbf/in}^2$$

The maximum stress is the principal stress. Since there is no longitudinal (axial) stress, the principal stresses in the shaft are

$$\tau_{\max} = \tfrac{1}{2}\sqrt{\sigma^2 + (2\tau)^2}$$
$$= \tfrac{1}{2}\sqrt{\left(5093 \frac{\text{lbf}}{\text{in}^2}\right)^2 + \left[(2)\left(1273 \frac{\text{lbf}}{\text{in}^2}\right)\right]^2}$$
$$= 2847 \text{ lbf/in}^2$$

$$\sigma_1, \sigma_2 = \tfrac{1}{2}\sigma \pm \tau_{\max}$$
$$= \left(\tfrac{1}{2}\right)\left(5093 \frac{\text{lbf}}{\text{in}^2}\right) \pm 2847 \frac{\text{lbf}}{\text{in}^2}$$
$$= 5394 \text{ lbf/in}^2, -301 \text{ lbf/in}^2$$

Use the distortion energy theory. The von Mises stress is

$$\sigma' = \sqrt{(\sigma_1)^2 + (\sigma_2)^2 - \sigma_1\sigma_2}$$

$$= \sqrt{\begin{array}{l}\left(5394 \frac{\text{lbf}}{\text{in}^2}\right)^2 + \left(-301 \frac{\text{lbf}}{\text{in}^2}\right)^2 \\ - \left(5394 \frac{\text{lbf}}{\text{in}^2}\right)\left(-301 \frac{\text{lbf}}{\text{in}^2}\right)\end{array}}$$

$$= 5551 \text{ lbf/in}^2$$

From the distortion energy theory, the factor of safety is

$$\text{FS} = \frac{S_{yt}}{\sigma'} = \frac{48{,}000 \frac{\text{lbf}}{\text{in}^2}}{5551 \frac{\text{lbf}}{\text{in}^2}}$$
$$= 8.65$$

(e) The shaft experiences alternating combined stresses. As such, both the reversed bending and shear must be considered.

The bending stress completely reverses itself each time the shaft turns half of a revolution. The mean bending stress is

$$\sigma_m = 0$$

The shaft torsion is constant. Therefore, the mean shear stress is

$$\tau_m = 1273 \text{ lbf/in}^2$$

From combined stress theory with these two mean stresses, the principal mean stresses are

$$\tau_{\max, m} = \tfrac{1}{2}\sqrt{\sigma_m^2 + (2\tau_m)^2}$$
$$= \tfrac{1}{2}\sqrt{0 + \left[(2)\left(1273 \frac{\text{lbf}}{\text{in}^2}\right)\right]^2}$$
$$= 1273 \text{ lbf/in}^2$$

$$\sigma_{m,1}, \sigma_{m,2} = \tfrac{1}{2}\sigma_m \pm \tau_{\max, m}$$
$$= \left(\tfrac{1}{2}\right)(0) \pm 1273 \frac{\text{lbf}}{\text{in}^2}$$
$$= +1273 \text{ lbf/in}^2, -1273 \text{ lbf/in}^2$$

The von Mises mean stress is

$$\sigma_m' = \sqrt{(\sigma_{m,1})^2 + (\sigma_{m,2})^2 - \sigma_{m,1}\sigma_{m,2}}$$

$$= \sqrt{\begin{array}{l}\left(1273 \frac{\text{lbf}}{\text{in}^2}\right)^2 + \left(-1273 \frac{\text{lbf}}{\text{in}^2}\right)^2 \\ - \left(1273 \frac{\text{lbf}}{\text{in}^2}\right)\left(-1273 \frac{\text{lbf}}{\text{in}^2}\right)\end{array}}$$

$$= 2205 \text{ lbf/in}^2$$

The bending stress varies from $+5093 \text{ lbf/in}^2$ (tension) to -5093 lbf/in^2 (compression). Therefore, the alternating bending stress is

$$\sigma_a = 5093 \text{ lbf/in}^2$$

The shear stress is constant. Therefore, the alternating shear stress is

$$\tau_a = 0$$

From combined stress theory with these two alternating stresses, the principal alternating stresses are

$$\tau_{\text{max},a} = \tfrac{1}{2}\sqrt{\sigma_a^2 + (2\tau_a)^2}$$

$$= \tfrac{1}{2}\sqrt{\left(5093\ \frac{\text{lbf}}{\text{in}^2}\right)^2 + (2)(0)^2}$$

$$= \left(\tfrac{1}{2}\right)\left(5093\ \frac{\text{lbf}}{\text{in}^2}\right)$$

$$\sigma_{a,1}, \sigma_{a,2} = \tfrac{1}{2}\sigma_a \pm \tau_{\text{max},a}$$

$$= \left(\tfrac{1}{2}\right)\left(5093\ \frac{\text{lbf}}{\text{in}^2}\right) \pm \left(\tfrac{1}{2}\right)\left(5093\ \frac{\text{lbf}}{\text{in}^2}\right)$$

$$= +5093\ \text{lbf/in}^2, 0\ \text{lbf/in}^2$$

The von Mises alternating stress is

$$\sigma_a' = \sqrt{(\sigma_{a,1})^2 + (\sigma_{a,2})^2 - \sigma_{a,1}\sigma_{a,2}}$$

$$= \sqrt{\left(5093\ \frac{\text{lbf}}{\text{in}^2}\right)^2 + (0)^2 - \left(5093\ \frac{\text{lbf}}{\text{in}^2}\right)\left(0\ \frac{\text{lbf}}{\text{in}^2}\right)}$$

$$= 5093\ \text{lbf/in}^2$$

These stresses are very low in comparison to the endurance and yield strengths. It is not necessary to draw the Goodman diagram.

The equivalent von Mises (Goodman diagram) stress is

$$\sigma_{\text{eq}}' = \sigma_a' + \left(\frac{S_e}{S_{ut}}\right)\sigma_m'$$

$$= 5093\ \frac{\text{lbf}}{\text{in}^2} + \left(\frac{35{,}000\ \dfrac{\text{lbf}}{\text{in}^2}}{59{,}000\ \dfrac{\text{lbf}}{\text{in}^2}}\right)\left(2205\ \frac{\text{lbf}}{\text{in}^2}\right)$$

$$= 6401\ \text{lbf/in}^2$$

The factor of safety is

$$\text{FS} = \frac{S_e}{\sigma_{\text{eq}}'} = \frac{35{,}000\ \dfrac{\text{lbf}}{\text{in}^2}}{6401\ \dfrac{\text{lbf}}{\text{in}^2}}$$

$$= 5.47$$

5 SOLUTION

(a) Proof testing should occur at the lowest temperature expected since the thermal shrinkage will cause additional longitudinal stress in the pipe. Test at 40°F.

(b) For 8 in schedule-40 pipe,

$$r_i = \frac{7.981\ \text{in}}{2} = 3.991\ \text{in}$$

$$r_o = \frac{8.625\ \text{in}}{2} = 4.312\ \text{in}$$

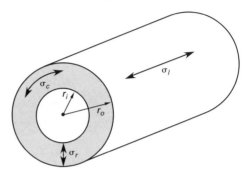

The unit strain caused by drawing the flanges together during installation at 80°F is

$$\epsilon_i = \frac{\delta}{L} = \frac{0.5\ \text{in}}{(100\ \text{ft})\left(12\ \dfrac{\text{in}}{\text{ft}}\right)}$$

$$= 4.167 \times 10^{-4}$$

The additional thermal strain at 40°F is

$$\epsilon_{\text{th}} = \alpha(T_1 - T_2)$$

$$= \left(6.5 \times 10^{-6}\ \frac{1}{{}^\circ\text{F}}\right)(80^\circ\text{F} - 40^\circ\text{F})$$

$$= 2.6 \times 10^{-4}$$

The total longitudinal stress at 40°F will be

$$\sigma_l = E\epsilon_{\text{total}} = E(\epsilon_i + \epsilon_{\text{th}})$$

$$= \left(3 \times 10^7\ \frac{\text{lbf}}{\text{in}^2}\right)(4.167 \times 10^{-4} + 2.6 \times 10^{-4})$$

$$= 20{,}301\ \text{lbf/in}^2$$

(c) The radial stress at the inner face is the pressurization.

$$r_{ri} = -p = -1000\ \text{lbf/in}^2 \quad \text{[compressive]}$$

Use Lamé's thick cylinder solutions to determine the circumferential stress at the inner face.

$$\sigma_{ci} = \frac{(r_o^2 + r_i^2)p}{r_o^2 - r_i^2}$$

$$= \frac{\left((3.991\ \text{in})^2 + (4.312\ \text{in})^2\right)\left(1000\ \dfrac{\text{lbf}}{\text{in}^2}\right)}{(4.312\ \text{in})^2 - (3.991\ \text{in})^2}$$

$$= 12{,}952\ \text{lbf/in}^2$$

Since there is no torsional stress, σ_l, σ_c, and σ_r are the principal stresses.

$$\sigma_1 = \sigma_l = 20{,}300 \text{ lbf/in}^2$$
$$\sigma_2 = \sigma_c = -1000 \text{ lbf/in}^2$$
$$\sigma_3 = \sigma_r = 12{,}952 \text{ lbf/in}^2$$

(d) Use the distortion energy theory for triaxial (three-dimensional) loading. The von Mises stress is

$$\sigma' = \sqrt{\left(\tfrac{1}{2}\right)\left[(\sigma_1 - \sigma_2)^2 + (\sigma_2 - \sigma_3)^2 + (\sigma_3 - \sigma_1)^2\right]}$$

$$= \sqrt{\left(\tfrac{1}{2}\right)\left[\begin{array}{l}\left(20{,}301 \dfrac{\text{lbf}}{\text{in}^2} - \left(-1000 \dfrac{\text{lbf}}{\text{in}^2}\right)\right)^2 \\[2mm] + \left(-1000 \dfrac{\text{lbf}}{\text{in}^2} - 12{,}952 \dfrac{\text{lbf}}{\text{in}^2}\right)^2 \\[2mm] + \left(12{,}952 \dfrac{\text{lbf}}{\text{in}^2} - 20{,}301 \dfrac{\text{lbf}}{\text{in}^2}\right)^2\end{array}\right]}$$

$$= 18{,}740 \text{ lbf/in}^2$$

The factor of safety is

$$\text{FS} = \frac{S_{yt}}{\sigma'} = \frac{30{,}000 \dfrac{\text{lbf}}{\text{in}^2}}{18{,}740 \dfrac{\text{lbf}}{\text{in}^2}}$$

$$= 1.6$$

Since $\sigma' < S_{yt}$ (18,740 psi $<$ 30,000 psi), the pipe will not yield and 1000 psig is a safe proof pressure. The factor of safety for ductile steel is probably okay.

(e) The pipe experiences temperature swings in both directions. By prestressing the pipe in tension, compressive stresses that would tend to buckle the pipeline at higher temperatures are eliminated.

6 SOLUTION

(a) The torque on the shaft is

$$T = \frac{\left(33{,}000 \dfrac{\text{ft-lbf}}{\text{hp-min}}\right) P}{2\pi n}$$

$$= \frac{\left(33{,}000 \dfrac{\text{ft-lbf}}{\text{hp-min}}\right)(25 \text{ hp})\left(12 \dfrac{\text{in}}{\text{ft}}\right)}{(2\pi)\left(1500 \dfrac{\text{rev}}{\text{min}}\right)}$$

$$= 1050 \text{ in-lbf}$$

Shafts fail in torsion (shear). Use the distortion energy theory to determine the yield stress in shear.

$$S_{ys} = 0.577 S_{yt} = (0.577)\left(86{,}000 \frac{\text{lbf}}{\text{in}^2}\right)$$

$$= 49{,}622 \text{ lbf/in}^2$$

Use the factor of safety on the material properties.

$$\tau_{\max} = \frac{S_{ys}}{\text{FS}} = \frac{49{,}622 \dfrac{\text{lbf}}{\text{in}^2}}{2}$$

$$= 24{,}811 \text{ lbf/in}^2$$

The diameter of the shaft is calculated from simple elastic theory.

$$d = \sqrt[3]{\frac{16T}{\pi \tau_{\max}}}$$

$$= \sqrt[3]{\frac{(16)(1050 \text{ in-lbf})}{\pi \left(24{,}811 \dfrac{\text{lbf}}{\text{in}^2}\right)}}$$

$$= 0.6 \text{ in}$$

(b) The common assumption is that of uniform wear.

The mean torque-carrying radius for the plates and disks is

$$r_m = \tfrac{1}{2}(r_o + r_i) = \tfrac{1}{2}\left[\left(\frac{6 \text{ in}}{2}\right) + \left(\frac{1 \text{ in}}{2}\right)\right]$$

$$= 1.75 \text{ in}$$

The mean tangential frictional force at the mean radius is

$$F_f = \frac{T}{r_m} = \frac{1050 \text{ in-lbf}}{1.75 \text{ in}}$$

$$= 600 \text{ lbf}$$

The frictional force can also be calculated from the normal force (i.e., the contact pressure). Let N_{plates} be the number of contact surfaces between the plates and disks.

$$F_f = fN = fpA = fp\left(\frac{\pi}{4}\right)d^2$$

$$600 \text{ lbf} = N_{\text{plates}}(0.1)\left(100 \frac{\text{lbf}}{\text{in}^2}\right)\left(\frac{\pi}{4}\right)$$
$$\times \left[(6 \text{ in})^2 - (1 \text{ in})^2\right]$$

$$N_{\text{plates}} = 2.18$$

The number of contact surfaces must be an even number. Since two surfaces would result in a contact surface greater than 100 lbf/in^2, four surfaces are required. This can be accomplished with three plates.

Adding 2 $(4-2)$ contact surfaces when only 0.18 $(2.18-2)$ is needed is an expensive overcapacity. In actual practice, some other design parameter, such as the clutch diameter or clutch material, would be adjusted to eliminate the overdesign that results here.

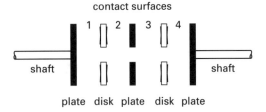

contact surfaces

plate   disk   plate   disk   plate

(c) The driving member turns at a constant speed. At the start of engagement, the driven member is not turning at all, and all of the input power is wasted. At the end of the engagement, the driven member is turning at full speed, and none of the input power is wasted. Since the acceleration is constant, during engagement, an average of one-half of the input power is wasted.

$$E_{\text{wasted}} = \frac{\left(\frac{1}{2}\right)(25 \text{ hp})\left(550 \frac{\text{ft-lbf}}{\text{hp-sec}}\right)(30 \text{ sec})}{778 \frac{\text{ft-lbf}}{\text{Btu}}}$$

$$= 265 \text{ Btu}$$

Assume all of this energy is absorbed by the plates. Disregard any heat transfer to the disks, clutch housing, or surrounding air and lubricants. The mass of the plates is

$$m = N_{\text{plates}}\rho V = N_{\text{plates}}\rho A t$$

$$= \frac{(3)\left(491 \frac{\text{lbm}}{\text{ft}^3}\right)\left(\frac{\pi}{4}\right)[(6 \text{ in})^2 - (1 \text{ in})^2](0.2 \text{ in})}{\left(12 \frac{\text{in}}{\text{ft}}\right)^3}$$

$$= 4.686 \text{ lbm}$$

The average temperature rise in the plates is

$$\Delta T = \frac{E_{\text{wasted}}}{mc_p}$$

$$= \frac{265 \text{ Btu}}{(4.686 \text{ lbm})\left(0.107 \frac{\text{Btu}}{\text{lbm-}°\text{F}}\right)}$$

$$= 529°\text{F}$$

No maximum temperature was stated for the clutch. However, this is probably an excessive temperature rise, particularly if the clutch material is traditional woven asbestos (maximum temperature approximately 500°F).

(d) There are several approaches that could be taken recommending against using a clutch plate as a disk-brake rotor. (1) Clutch plates, being solid, do not have the perforated ventilation holes or "honeycomb" structure that disk-brake rotors have to dissipate heat quickly. (2) Clutch plates do not have the rigidity that disk-brake rotors have, and the engagement pressure would not be uniform. (3) The central area of the clutch plate would not be used, thus contributing to higher material costs and somewhat higher rotational inertia. Disk-brake rotors typically concentrate their mass at the outer edges, where the calipers grip them.

7   SOLUTION

(a) The original production rate is given per month. There are more than 4 weeks per month. The production rate per week using the original manual method is

$$\frac{\left(12{,}000 \frac{\text{units}}{\text{month}}\right)\left(12 \frac{\text{months}}{\text{yr}}\right)}{52 \frac{\text{wk}}{\text{yr}}} = 2769 \text{ units/wk}$$

The interest rate given is a nominal interest rate, since compounding is semiannual. The nominal interest rate can be converted to an effective rate per year (10.25%) without too much difficulty. However, the fact that machines B and C become available in 6 months is problematic. It is easiest to work this problem in terms of periods of 6 months. This allows a standard interest table to be used.

The effective annual interest rate per semiannual period is

$$i = \frac{r}{k} = \frac{0.10}{2} = 0.05 \quad (5\%)$$

The semiannual (26 weeks) cost of one employee is

$$(26 \text{ wk})\left(5 \frac{\text{days}}{\text{wk}}\right)\left(8 \frac{\text{hr}}{\text{day}}\right)\left(\frac{\$15}{\text{hr}}\right) = \$15{,}600$$

The current semiannual expense associated for all five employees is

$$A = (5)(\$15{,}600) = \$78{,}000$$

The horizon is 12 compounding periods (6 years). Use the factor tables to determine the uniform series present worth factor. The present worth of the alternative to manufacture the product manually is

$$P_{\text{O}} = A(P/A, i\%, n) = (\$78{,}000)(P/A, 5\%, 12)$$
$$= (\$78{,}000)(8.8633)$$
$$= \$691{,}337$$

Now, evaluate the alternative machines one by one.

Machine A

The average number of good products manufactured each hour is

$$(\text{production})(\text{availability})(\text{yield})$$
$$= \left(100 \ \frac{\text{units}}{\text{hr}}\right)(0.90)(0.88)$$
$$= 79.2 \ \text{units/hr}$$

The number of hours needed to sustain the original rate is

$$\frac{2769 \ \dfrac{\text{units}}{\text{wk}}}{79.2 \ \dfrac{\text{units}}{\text{hr}}} = 34.96 \ \text{hr/wk}$$

Since there are 40 hours in a normal work week, machine A has the needed capacity without requiring overtime.

Recalling that the compounding period is half a year, the expenses associated with machine A are:

- initial cost: $200,000 at $t = 0$

- labor for first year: $78,000 at $t = 1$ and $t = 2$

- labor for subsequent years:
  $15,600 at $t = 3$ through 12

- maintenance:
  $10,000/2 = \$5,000$ at $t = 3$ through 12

The present worth of the expenses is
$$
\begin{aligned}
P_A = \ & \$200{,}000 \\
& + (\$78{,}000)[(P/F, 5\%, 1) + (P/F, 5\%, 2)] \\
& + (\$15{,}600 + \$5000) \\
& \quad \times [(P/A, 5\%, 12) - (P/A, 5\%, 2)] \\
= \ & \$200{,}000 \\
& + (\$78{,}000)(0.9524 + 0.9070) \\
& + (\$15{,}600 + \$5000)(8.8633 - 1.8594) \\
= \ & \$489{,}314
\end{aligned}
$$

Machine B

The average number of good products manufactured each hour is

$$(\text{production})(\text{availability})(\text{yield})$$
$$= \left(87.5 \ \frac{\text{units}}{\text{hr}}\right)(0.80)(0.92)$$
$$= 64.4 \ \text{units/hr}$$

The number of hours needed to sustain the original rate is

$$\frac{2769 \ \dfrac{\text{units}}{\text{wk}}}{64.4 \ \dfrac{\text{units}}{\text{hr}}} = 43.00 \ \text{hrs/wk}$$

Machine B does not have the needed capacity without overtime. Three hours of overtime will be required per week. The semiannual cost of this overtime is

$$(26 \ \text{wk})\left(3 \ \frac{\text{hr}}{\text{wk}}\right)(1.5)\left(\frac{\$15}{\text{hr}}\right) = \$1755$$

The expenses associated with machine B are:

- initial cost: $175,000 at $t = 0$

- labor for first 6-month period: $78,000 at $t = 1$

- labor for subsequent periods:
  $15,600 at $t = 2$ through 12

- overtime: $1755 at $t = 2$ through 12

- maintenance:
  $15,000/2 = \$7{,}500$ at $t = 2$ through 12

The present worth of the expenses is

$$
\begin{aligned}
P_B = \ & \$175{,}000 \\
& + (\$78{,}000)(P/F, 5\%, 1) \\
& + (\$15{,}600 + \$1755 + \$7500) \\
& \quad \times [(P/A, 5\%, 12) - (P/A, 5\%, 1)] \\
= \ & \$175{,}000 \\
& + (\$78{,}000)(0.9524) \\
& + (\$15{,}600 + \$1755 + \$7500)(8.8633 - 0.9524) \\
= \ & \$445{,}913
\end{aligned}
$$

Machine C

The average number of good products manufactured each hour is

$$(\text{production})(\text{availability})(\text{yield})$$
$$= \left(82 \ \frac{\text{units}}{\text{hr}}\right)(0.75)(0.95)$$
$$= 58.43 \ \text{units/hr}$$

The number of hours needed to sustain the original rate is

$$\frac{2769 \ \dfrac{\text{units}}{\text{wk}}}{58.43 \ \dfrac{\text{units}}{\text{hr}}} = 47.4 \ \text{hr/wk}$$

Machine C does not have the needed capacity without overtime. 7.4 hours of overtime will be required per week. This is in excess of the 4-hour limit, so one 8-hour swingshift will be needed each week. The semiannual cost of this swingshift is

$$(26 \ \text{wk})\left(8 \ \frac{\text{hr}}{\text{wk}}\right)(1.25)\left(\frac{\$15}{\text{hr}}\right) = \$3900$$

The expenses associated with machine C are:

- initial cost: $150,000 at $t = 0$

- labor for first 6-month period: $78,000 at $t = 1$

- labor for subsequent periods:
  $15,600 at $t = 2$ through 12

- swingshift: $3900 at $t = 2$ through 12

- maintenance:
  $17,000/2 = \$8,500$ at $t = 2$ through 12

The present worth of the expenses is

$$P_{\text{C}} = \$150,000$$
$$+ (\$78,000)(P/F, 5\%, 1)$$
$$+ (\$15,600 + \$3900 + \$8500)$$
$$\times [(P/A, 5\%, 12) - (P/A, 5\%, 1)]$$
$$= \$150,000$$
$$+ (\$78,000)(0.9524)$$
$$+ (\$15,600 + \$3900 + \$8500)(8.8633 - 0.9524)$$
$$= \$445,792$$

Based on these costs alone, machine C has the lowest present worth of expenses and should be selected.

(b) The present worths of machines B and C are so close as to be essentially identical. That means that minor errors in the cost data could shift the decision one way or the other. Due to unknowns regarding future costs and the value of money, these two alternatives are probably equivalent. A decision should take nonquantifiable factors (e.g., environmental friendliness, relationship with the machine vendor, union and labor issues) into consideration.

8 SOLUTION

(a) The discharge from an orifice under pressure can be calculated from Toricelli's equation using the discharge coefficient (coefficient of discharge): $Q = C_d A_o \sqrt{2gh}$. However, a ½ in orifice with coefficient of discharge of 0.75 corresponds to the "standard" sprinkler, and there is no need to use purely theoretical methods. For a standard sprinkler, the discharge constant is 5.6.

Only the normal pressure is used with the last sprinkler.

$$Q = 5.6\sqrt{p} = 5.6\sqrt{10 \text{ psig}}$$
$$= 17.7 \text{ gal/min}$$

(b) Disregarding the velocity head, the normal pressure at the second sprinkler is the normal pressure at the end sprinkler plus the friction loss between the two sprinklers. The friction loss can be found from the Hazen-Williams equation, or it can be found from a friction loss chart (the typical method). From a chart, the loss at 17.7 gal/min and 1 in pipe is easily read at approximately 0.1 psig/ft.

The normal pressure at the second sprinkler is

$$p = 10 \text{ psig} + (10 \text{ ft})\left(0.1 \frac{\text{psig}}{\text{ft}}\right) = 11 \text{ psig}$$

The discharge is

$$Q = 5.6\sqrt{p} = 5.6\sqrt{11 \text{ psig}}$$
$$= 18.6 \text{ gal/min}$$

(c) From a pipe table, the flow area of a 1 in diameter schedule-40 pipe is 0.0060 ft^2.

Velocity pressure is not included in the discharge calculation for the end sprinkler.

The velocity in the pipe at the second sprinkler is

$$v = \frac{\dot{V}}{A} = \frac{\left(17.7 \frac{\text{gal}}{\text{min}}\right)\left(0.002228 \frac{\text{ft}^3\text{-min}}{\text{sec-gal}}\right)}{0.0060 \text{ ft}^2}$$
$$= 6.57 \text{ ft/sec}$$

The velocity pressure for the second sprinkler is

$$p_v = \frac{0.433 \, v^2}{2g}$$
$$= \frac{\left(0.433 \frac{\frac{\text{lbf}}{\text{in}^2}}{\text{ft}}\right)\left(6.57 \frac{\text{ft}}{\text{sec}}\right)^2}{(2)\left(32.2 \frac{\text{ft}}{\text{sec}^2}\right)}$$
$$= 0.29 \text{ lbf/in}^2$$

The available normal pressure at the second sprinkler is

$$p = 11 \text{ psig} - 0.29 \text{ psig} = 10.71 \text{ psig}$$

The discharge from the second sprinkler adjusted for velocity pressure is

$$Q = 5.6\sqrt{p} = 5.6\sqrt{10.71 \text{ psig}}$$
$$= 18.3 \text{ gal/min}$$

The total flow between the second and third sprinklers is

$$17.7 \frac{\text{gal}}{\text{min}} + 18.3 \frac{\text{gal}}{\text{min}} = 36.0 \text{ gal/min}$$

The velocity in the pipe at the third sprinkler is

$$v = \frac{\dot{V}}{A} = \frac{\left(36.0 \, \frac{\text{gal}}{\text{min}}\right)\left(0.002228 \, \frac{\text{ft}^3\text{-min}}{\text{sec-gal}}\right)}{0.0060 \, \text{ft}^2}$$
$$= 13.37 \, \text{ft/sec}$$

The velocity pressure for the second sprinkler is

$$p_v = \frac{0.433 \, v^2}{2g}$$
$$= \frac{\left(0.433 \, \frac{\text{lbf}}{\text{in}^2}\right)\left(13.37 \, \frac{\text{ft}}{\text{sec}}\right)^2}{(2)\left(32.2 \, \frac{\text{ft}}{\text{sec}^2}\right)}$$
$$= 1.20 \, \text{lbf/in}^2$$

The friction loss with 36 gal/min flowing in a 1 in pipe is approximately 0.38 lbf/in^2/ft. The friction loss in the 10 ft of pipe between sprinklers 2 and 3 is

$$\left(0.38 \, \frac{\frac{\text{lbf}}{\text{in}^2}}{\text{ft}}\right)(10 \, \text{ft}) = 3.8 \, \text{lbf/in}^2$$

The available normal pressure at the second sprinkler is

$$p = 10.7 \, \text{psig} + 3.8 \, \frac{\text{lbf}}{\text{in}^2} - 1.20 \, \text{psig} = 13.3 \, \text{psig}$$

The discharge from the second sprinkler adjusted for velocity pressure is

$$Q = 5.6\sqrt{p} = 5.6\sqrt{13.3 \, \text{psig}}$$
$$= 20.4 \, \text{gal/min}$$

(d) A minimum-sized pump requires the pump's overload capacity to be used, which is 150% of rated capacity. Therefore, the rated capacity is

$$\frac{1300 \, \frac{\text{gal}}{\text{min}}}{1.5} = 867 \, \text{gal/min}$$

The closest standard pump size is 1000 gal/min.

9 SOLUTION

(a) The characteristic equation is derived from the homogeneous differential equation.

$$s^2 + 5s + 16 = 0$$

(b) Traditional analysis shows that the homogeneous linear differential equation can be put in the following format:

$$y'' + 2\zeta\omega y' + \omega^2 y = 0$$

In this problem, $\omega^2 = 16$. Therefore,

$$\omega = \sqrt{16} = 4 \, \text{rad/sec}$$

(c) Since $2\zeta\omega = 5$,

$$\zeta = \frac{5}{2\omega} = \frac{5}{(2)(4)} = 5/8$$

(d) The damped natural frequency is

$$\omega_d = \omega\sqrt{1 - \zeta^2}$$
$$= \left(4 \, \frac{\text{rad}}{\text{sec}}\right)\sqrt{1 - \left(\frac{5}{8}\right)^2}$$
$$= 3.12 \, \text{rad/sec}$$

(e) The time constant is

$$\tau = \frac{1}{\zeta\omega} = \frac{1}{\left(\frac{5}{8}\right)\left(4 \, \frac{\text{rad}}{\text{sec}}\right)}$$
$$= 2/5 \, \text{sec}$$

10 SOLUTION

(a) In the absence of specific information about the location of the vena contracta, the standard configuration is 1 pipe diameter upstream and $1/2$ pipe diameter downstream.

(b) Tap holes should be at least 5 to 10 pipe diameters beyond any source of turbulence.

(c) The ratio of orifice area to upstream area is

$$\frac{A_o}{A_1} = \left(\frac{D_o}{D_1}\right)^2 = \left(\tfrac{1}{2}\right)^2 = 0.25$$

From a graph of flow coefficients for orifice plates for this area ratio, the flow coefficient does not change after a Reynolds number of slightly less than 10^5.

(d) The inside diameter for 12 in pipe is

$$D_i = 11.938 \, \text{in}$$

The diameter and area of the orifice are

$$D_o = \frac{11.938 \text{ in}}{(2)\left(12 \frac{\text{in}}{\text{ft}}\right)} = 0.497 \text{ ft}$$

$$A_o = \left(\frac{\pi}{4}\right) D_o^2 = \left(\frac{\pi}{4}\right)(0.497 \text{ ft})^2 = 0.194 \text{ ft}^2$$

The flow velocity is

$$\text{v} = \frac{\dot{V}}{A} = \frac{\left(1400 \frac{\text{gal}}{\text{min}}\right)\left(0.002228 \frac{\text{ft}^3\text{-min}}{\text{sec-gal}}\right)}{0.194 \text{ ft}^2}$$

$$= 16.08 \text{ ft/sec}$$

The kinematic viscosity of 60°F water is

$$\nu = 1.217 \times 10^{-5} \text{ ft}^2/\text{sec}$$

The Reynolds number is

$$\text{Re} = \frac{D\text{v}}{\nu} = \frac{(0.497 \text{ ft})\left(16.08 \frac{\text{ft}}{\text{sec}}\right)}{1.217 \times 10^{-5} \frac{\text{ft}^2}{\text{sec}}} = 6.57 \times 10^5$$

From a graph of flow coefficients for orifice plates for an area ratio of 0.25 and this Reynolds number, the flow coefficient is approximately 0.625.

(e) The permanent pressure drop coefficient (in multiples of the orifice velocity head) across the orifice plate is

$$K = \frac{1 - \beta^2}{C_f^2} = \frac{1 - (0.5)^2}{(0.625)^2} = 1.92$$

The permanent pressure drop is

$$\Delta p = \rho \left(\frac{g}{g_c}\right) K h_\text{v} = \gamma K \left(\frac{\text{v}^2}{2g}\right)$$

$$= \frac{\left(62.37 \frac{\text{lbf}}{\text{ft}^3}\right)(1.92)\left(16.08 \frac{\text{ft}}{\text{sec}}\right)^2}{(2)\left(32.2 \frac{\text{ft}}{\text{sec}^2}\right)}$$

$$= 480.8 \text{ lbf/ft}^2 \quad (3.38 \text{ lbf/in}^2)$$

(f) Assume $C_f = 0.62$.

$$\dot{V} = C_f A_o \sqrt{\frac{2g_c(p_1 - p_2)}{\rho}}$$

$$= (0.62)(0.194 \text{ ft}^2)\sqrt{\frac{(2)\left(32.2 \frac{\text{ft-lbm}}{\text{lbf-sec}^2}\right) \times \left(6 \frac{\text{lbf}}{\text{in}^2}\right)\left(144 \frac{\text{in}^2}{\text{ft}^2}\right)}{62.37 \text{ lbm}^3}}$$

$$= 3.593 \text{ ft}^3/\text{sec} \quad (1612 \text{ gal/min})$$

Check the assumed value of C_f. The flow velocity is

$$\text{v} = \frac{\dot{V}}{A} = \frac{3.593 \frac{\text{ft}^3}{\text{sec}}}{0.194 \text{ ft}^2} = 18.52 \text{ ft/sec}$$

The Reynolds number is

$$\text{Re} = \frac{D\text{v}}{\nu} = \frac{(0.497 \text{ ft})\left(18.52 \frac{\text{ft}}{\text{sec}}\right)}{1.217 \times 10^{-5} \frac{\text{ft}^2}{\text{sec}}} = 7.56 \times 10^5$$

From a graph of flow coefficients for orifice plates for an area ratio of 0.25 and this Reynolds number, the flow coefficient is approximately 0.625 (which is OK).

(g) Assume $C_f = 0.98$.

$$\dot{V} = C_f A_2 \sqrt{\frac{2g_c(p_1 - p_2)}{\rho}}$$

$$= (0.98)(0.194 \text{ ft}^2)\sqrt{\frac{(2)\left(32.2 \frac{\text{ft-lbm}}{\text{lbf-sec}^2}\right) \times \left(6 \frac{\text{lbf}}{\text{in}^2}\right)\left(144 \frac{\text{in}^2}{\text{ft}^2}\right)}{62.37 \frac{\text{lbm}}{\text{ft}^3}}}$$

$$= 5.679 \text{ ft}^3/\text{sec} \quad (2549 \text{ gal/min})$$

Check the assumed value of C_f. The flow velocity is

$$\text{v} = \frac{\dot{V}}{A} = \frac{5.679 \frac{\text{ft}^3}{\text{sec}}}{0.194 \text{ ft}^2} = 29.27 \text{ ft/sec}$$

The Reynolds number is

$$\text{Re} = \frac{D\text{v}}{\nu} = \frac{(0.497 \text{ ft})\left(29.27 \frac{\text{ft}}{\text{sec}}\right)}{1.217 \times 10^{-5} \frac{\text{ft}^2}{\text{sec}}} = 1.20 \times 10^6$$

From a graph of flow coefficients for venturi meters for this Reynolds number, the flow coefficient is approximately 0.99.

$$\dot{V} = C_f A_2 \sqrt{\frac{2g_c(p_1 - p_2)}{\rho}}$$

$$= (0.99)(0.194 \text{ ft}^2)\sqrt{\frac{(2)\left(32.2 \frac{\text{ft-lbm}}{\text{lbf-sec}^2}\right) \times \left(6 \frac{\text{lbf}}{\text{in}^2}\right)\left(144 \frac{\text{in}^2}{\text{ft}^2}\right)}{62.37 \frac{\text{lbm}}{\text{ft}^3}}}$$

$$= 5.737 \text{ ft}^3/\text{sec} \quad (2575 \text{ gal/min})$$

(h) The bulk modulus of water at 60°F is 311×10^3 psi. The modulus of elasticity of steel is approximately 30×10^6 psi. (An argument for 29×10^6 psi for ductile steel could also be made.)

The wall thickness and inside diameter for 12 in pipe are

$$t = 0.406 \text{ in}$$
$$D_i = 11.938 \text{ in}$$

The modulus of elasticity used in calculating the speed of sound is

$$E = \frac{E_{\text{water}} t_{\text{pipe}} E_{\text{pipe}}}{t_{\text{pipe}} E_{\text{pipe}} + D_{\text{pipe}} E_{\text{water}}}$$

$$= \frac{\left(311 \times 10^3 \, \dfrac{\text{lbf}}{\text{in}^2}\right) (0.406 \text{ in}) \left(30 \times 10^6 \, \dfrac{\text{lbf}}{\text{in}^2}\right)}{(0.406 \text{ in}) \left(30 \times 10^6 \, \dfrac{\text{lbf}}{\text{in}^2}\right)}$$
$$+ (11.938 \text{ in}) \left(311 \times 10^3 \, \dfrac{\text{lbf}}{\text{in}^2}\right)$$

$$= 2.383 \times 10^5 \text{ lbf/in}^2$$

The speed of sound in the pipe is

$$a = \sqrt{\frac{E}{\rho}}$$

$$= \sqrt{\frac{\left(2.383 \times 10^5 \, \dfrac{\text{lbf}}{\text{in}^2}\right) \left(144 \, \dfrac{\text{in}^2}{\text{ft}^2}\right)}{62.37 \, \dfrac{\text{lbm}}{\text{ft}^3}}}$$

$$= 741.7 \text{ ft/sec}$$

Assuming the water comes to a dead stop, the pressure increase is

$$\Delta p = \frac{\rho a \Delta \text{v}}{g_c}$$

$$= \frac{\left(62.37 \, \dfrac{\text{lbm}}{\text{ft}^3}\right) \left(741.7 \, \dfrac{\text{ft}}{\text{sec}}\right) \left(20 \, \dfrac{\text{ft}}{\text{sec}}\right)}{32.2 \, \dfrac{\text{ft-lbm}}{\text{lbf-sec}^2}}$$

$$= 2.87 \times 10^4 \text{ lbf/ft}^2 \quad (199 \text{ lbf/in}^2)$$

(i) The hoop stress is one of the principal stresses in the pipe. It is appropriate to use a thin-wall model with this pipe.

$$\sigma_h = \frac{(p_{\text{static}} + \Delta p) r}{t}$$

$$= \frac{\left(200 \, \dfrac{\text{lbf}}{\text{in}^2} + 500 \, \dfrac{\text{lbf}}{\text{in}^2}\right) \left(\dfrac{11.938 \text{ in}}{2}\right)}{0.406 \text{ in}}$$

$$= 10,291 \text{ lbf/in}^2 \quad (10.3 \text{ kips/in}^2)$$

Solutions Part 2

11 SOLUTION

(a) The viscosity given in centistokes is actually the kinematic viscosity. Convert the units.

$$\nu = (1.13 \text{ cS}) \left(1.0764 \times 10^{-5} \frac{\text{ft}^2}{\text{sec-cS}} \right)$$
$$= 1.216 \times 10^{-5} \text{ ft}^2/\text{sec}$$

(This corresponds to the value of 1.217×10^{-5} ft^2/sec obtained from a table of water properties.)

The answer is A.

(b) The inside pipe diameter is

$$D_i = \frac{6.065 \text{ in}}{12 \frac{\text{in}}{\text{ft}}} = 0.5054 \text{ ft}$$

The pipe flow area is

$$A = \left(\frac{\pi}{4} \right) D^2 = \left(\frac{\pi}{4} \right) (0.5054 \text{ ft})^2 = 0.2006 \text{ ft}^2$$

The velocity is

$$\text{v} = \frac{\dot{V}}{A} = \frac{\left(400 \frac{\text{gal}}{\text{min}} \right) \left(0.002228 \frac{\text{ft}^2\text{-min}}{\text{sec-gal}} \right)}{0.2006 \text{ ft}^2}$$
$$= 4.443 \text{ ft/sec} \quad (4.4 \text{ ft/sec})$$

The answer is B.

(c) The Reynolds number is

$$\text{Re} = \frac{D\text{v}}{\nu} = \frac{(0.5054 \text{ ft}) \left(4.443 \frac{\text{ft}}{\text{sec}} \right)}{1.216 \times 10^{-5} \frac{\text{ft}^2}{\text{sec}}}$$
$$= 1.85 \times 10^5 \quad (18,500)$$

The answer is D.

(d) From the Moody friction factor chart for Re $= 1.84 \times 10^5$ and $\epsilon/D = 0.001$ (given), read

$$f = 0.021$$

The answer is C.

(e) The main friction loss in the line is

$$h_{f,\text{line}} = \frac{fL\text{v}^2}{2Dg}$$
$$= \frac{(0.021)(910 \text{ ft}) \left(4.443 \frac{\text{ft}}{\text{sec}} \right)^2}{(2)(0.5054 \text{ ft}) \left(32.2 \frac{\text{ft}}{\text{sec}^2} \right)}$$
$$= 11.59 \text{ ft}$$

The minor loss is given as 1.15 ft.

$$h_{f,\text{total}} = 11.59 + 1.15 = 12.74 \text{ ft} \quad (12.7 \text{ ft})$$

The answer is D.

(f) Disregarding velocity head, and since the pump inlet pressure is the same as the tank inlet pressure, the total head that the pump produces is

$$h_{\text{total}} = h_f + h_{\text{minor}} + h_z$$
$$= 33.1 \text{ ft} + 1.15 \text{ ft} + 806 \text{ ft}$$
$$= 840.25 \text{ ft} \quad (840 \text{ ft})$$

The answer is C.

(g) The density of 60°F water is 62.37 lbm/ft^3.

$$\dot{m} = \rho \dot{V} = \rho \text{v} A$$
$$= \left(62.37 \frac{\text{lbm}}{\text{ft}^3} \right) \left(6.54 \frac{\text{ft}}{\text{sec}} \right) \left(\frac{\pi}{4} \right) \left(\frac{5.000 \text{ in}}{12 \frac{\text{in}}{\text{ft}}} \right)^2$$
$$= 55.62 \text{ lbm/sec} \quad (55.6 \text{ lbm/sec})$$

The answer is C.

(h) Neglecting velocity head, the power required is

$$P = \frac{\dot{m}h_{\text{total}}}{\eta_{\text{pump}}\left(550\ \frac{\text{ft-lbf}}{\text{hp-sec}}\right)}\left(\frac{g}{g_c}\right)$$

$$= \left(\frac{\left(55.62\ \frac{\text{lbm}}{\text{sec}}\right)(840.25\ \text{ft})}{(0.70)\left(550\ \frac{\text{ft-lbf}}{\text{hp-sec}}\right)}\right)\left(\frac{32.2\ \frac{\text{ft}}{\text{sec}^2}}{32.2\ \frac{\text{ft-lbm}}{\text{lbf-sec}^2}}\right)$$

$$= 121.4\ \text{hp}\quad(120\ \text{hp})$$

The answer is C.

(i) The speed of sound in the water is

$$c = \sqrt{\frac{Eg_c}{\rho}}$$

$$= \sqrt{\frac{\left(320{,}000\ \frac{\text{lbf}}{\text{in}^2}\right)\left(144\ \frac{\text{in}^2}{\text{ft}^2}\right)\left(32.2\ \frac{\text{ft-lbm}}{\text{lbf-sec}^2}\right)}{62.37\ \frac{\text{lbm}}{\text{ft}^2}}}$$

$$= 4877\ \text{ft/sec}\quad(4900\ \text{ft/sec})$$

The answer is D.

(j) The static pressure at the pump outlet is

$$p_{\text{static}} = p_{\text{tank}} + p_z = p_{\text{tank}} + \gamma\Delta z$$

$$= p_{\text{tank}} + \rho\left(\frac{g}{g_c}\right)z$$

$$= 60\ \text{psig} + \frac{\left(62.37\ \frac{\text{lbm}}{\text{ft}^3}\right)\left(\frac{32.2\ \frac{\text{ft}}{\text{sec}^2}}{32.2\ \frac{\text{ft-lbm}}{\text{lbf-sec}^2}}\right)(806\ \text{ft})}{144\ \frac{\text{in}^2}{\text{ft}^2}}$$

$$= 409\ \text{psig}\quad(410\ \text{psig})$$

The answer is C.

12 SOLUTION

(a) $\quad \Delta E_k = \dfrac{v_B^2 - v_A^2}{2g_c}$

$$= \frac{\left(2161\ \frac{\text{ft}}{\text{sec}}\right)^2 - \left(190\ \frac{\text{ft}}{\text{sec}}\right)^2}{(2)\left(32.2\ \frac{\text{ft-lbm}}{\text{lbf-sec}^2}\right)\left(778\ \frac{\text{ft-lbf}}{\text{Btu}}\right)}$$

$$= 92.5\ \text{Btu/lbm}\quad(93\ \text{Btu/lbm})$$

The answer is B.

(b) $\quad h = u + \dfrac{pv}{J} = u + \dfrac{p}{J\rho}$

$$\Delta h = h_B - h_A$$

$$= u_B - u_A + \frac{\dfrac{p_B}{\rho_B} - \dfrac{p_A}{\rho_A}}{J}$$

$$= 53.1\ \frac{\text{Btu}}{\text{lbm}} - 119.1\ \frac{\text{Btu}}{\text{lbm}}$$

$$+ \frac{\left(\dfrac{5.85\ \frac{\text{lbf}}{\text{in}^2}}{0.0508\ \frac{\text{lbm}}{\text{ft}^3}} - \dfrac{98.4\ \frac{\text{lbf}}{\text{in}^2}}{0.381\ \frac{\text{lbm}}{\text{ft}^3}}\right)\left(144\ \frac{\text{in}^2}{\text{ft}^2}\right)}{778\ \frac{\text{ft-lbf}}{\text{Btu}}}$$

$$= -92.5\ \text{Btu/lbm}\quad(-93\ \text{Btu/lbm})$$

The answer is A.

(c) $\quad \text{flow energy} = \dfrac{pv}{J} = \dfrac{p}{J\rho}$

$$\Delta\text{flow energy} = \frac{\dfrac{p_B}{\rho_B} - \dfrac{p_A}{\rho_A}}{J}$$

$$= \frac{\left(\dfrac{5.85\ \frac{\text{lbf}}{\text{in}^2}}{0.0508\ \frac{\text{lbm}}{\text{ft}^3}} - \dfrac{98.4\ \frac{\text{lbf}}{\text{in}^2}}{0.381\ \frac{\text{lbm}}{\text{ft}^3}}\right)\left(144\ \frac{\text{in}^2}{\text{ft}^2}\right)}{778\ \frac{\text{ft-lbf}}{\text{Btu}}}$$

$$= -26.5\ \text{Btu/lbm}\quad(-27\ \text{Btu/lbm})$$

The answer is B.

(d)
$$E_p = z\left(\frac{g}{g_c}\right)$$

$$\Delta E_p = (z_B - z_A)\left(\frac{g}{g_c}\right)$$

$$= (2\text{ ft} - 0)\left(\frac{32.2\,\dfrac{\text{ft}}{\text{sec}^2}}{32.2\,\dfrac{\text{ft-lbm}}{\text{lbf-sec}^2}}\right)$$

$$= 2\text{ ft-lbf/lbm}$$

The answer is D.

(e) Use the properties at point A.

$$\dot{m} = \rho v A$$

$$= \left(0.381\,\frac{\text{lbm}}{\text{ft}^3}\right)\left(190\,\frac{\text{ft}}{\text{sec}}\right)\left(\frac{\pi}{4}\right)\left(\frac{5\text{ in}}{12\,\dfrac{\text{in}}{\text{ft}}}\right)^2$$

$$= 9.87\text{ lbm/sec}\quad(10\text{ lbm/sec})$$

The answer is B.

(f) The first law of thermodynamics for an open system is

$$q = \Delta h + \Delta E_k + \Delta E_p + W$$

$$= -92.5\,\frac{\text{Btu}}{\text{lbm}} + 92.5\,\frac{\text{Btu}}{\text{lbm}} + \frac{2\,\dfrac{\text{ft-lbf}}{\text{lbm}}}{778\,\dfrac{\text{ft-lbf}}{\text{Btu}}} + 0$$

$$= 0.00257\text{ Btu/lbm}$$

Notice that the enthalpy and kinetic energy differences cancel.

$$Q = \dot{m}q$$

$$= \left(9.87\,\frac{\text{lbm}}{\text{sec}}\right)\left(0.00257\,\frac{\text{Btu}}{\text{lbm}}\right)$$

$$= 0.0254\text{ Btu/sec}\quad(0.03\text{ Btu/sec})\quad[\text{gain}]$$

The answer is A.

(g)
$$W = Q - \Delta h - \Delta E_k - \Delta E_p$$

However, $\Delta h + \Delta E_k = 0$, so

$$W = Q - \Delta E_p = \dot{m}(q - \Delta E_p)$$

$$= \left(9.87\,\frac{\text{lbm}}{\text{sec}}\right)\left[\left(20\,\frac{\text{Btu}}{\text{lbm}}\right)\left(778\,\frac{\text{ft-lbf}}{\text{Btu}}\right) - 2\,\frac{\text{ft-lbf}}{\text{lbm}}\right]$$

$$= 1.536\times10^5\text{ ft-lbf/sec}\quad(150{,}000\text{ ft-lbf/sec})$$

The answer is D.

(h) Since the fluid is ideal,

$$c_p = \frac{R}{J} + c_p = \frac{52.9\,\dfrac{\text{ft-lbf}}{\text{lbm-}°\text{F}}}{778\,\dfrac{\text{ft-lbf}}{\text{Btu}}} + 0.170\,\frac{\text{Btu}}{\text{lbm-}°\text{F}}$$

$$= 0.238\text{ Btu/lbm-}°\text{F}$$

The answer is C.

(i) For a reversible adiabatic process, the polytropic exponent is equal to the ratio of specific heats.

$$n = k = \frac{c_p}{c_v} = \frac{0.238}{0.170}$$

$$= 1.40$$

The answer is D.

(j) Since the internal energy decreases, the temperature must also change. Choice (A) is false. Since enthalpy decreases, choice (B) is false. Since the pressure decreases, choice (C) is false. All data given fit an expansion process.

The answer is D.

13 SOLUTION

(a) Gather enthalpy data.

$$s_1 = 1.6751\text{ Btu/lbm-}°\text{R}$$

$$h_1 = 1368.3\text{ Btu/lbm}$$

$$s_2 = 1.6751\text{ Btu/lbm-}°\text{R}$$

$$x_2 = 0.95925$$

$$h_2 = 1117.5\text{ Btu/lbm}\quad[\text{isentropic}]$$

$$h_2' = 1368.3\,\frac{\text{Btu}}{\text{lbm}}$$

$$\quad - (0.87)\left(1368.3\,\frac{\text{Btu}}{\text{lbm}} - 1117.5\,\frac{\text{Btu}}{\text{lbm}}\right)$$

$$= 1150.1\text{ Btu/lbm}$$

$$s_3 = 1.6751\text{ Btu/lbm-}°\text{R}$$

$$x_3 = 0.92514$$

$$h_3 = 1069.8\text{ Btu/lbm}\quad[\text{isentropic}]$$

$$h_3' = 1368.3\,\frac{\text{Btu}}{\text{lbm}}$$

$$\quad - (0.87)\left(1368.3\,\frac{\text{Btu}}{\text{lbm}} - 1069.8\,\frac{\text{Btu}}{\text{lbm}}\right)$$

$$= 1108.6\text{ Btu/lbm}$$

$s_4 = 1.6751$ Btu/lbm-°R

$x_4 = 0.83585$

$h_4 = 935.7$ Btu/lbm   [isentropic]

$h'_4 = 1368.3 \dfrac{\text{Btu}}{\text{lbm}} - (0.87)\left(1368.3 \dfrac{\text{Btu}}{\text{lbm}} - 935.7 \dfrac{\text{Btu}}{\text{lbm}}\right)$

$\quad = 991.9$ Btu/lbm

$h_7 = 1156.4$ Btu/lbm

$h'_2 = 1150.1$ Btu/lbm   (1150 Btu/lbm)

The answer is B.

(b) Assuming a generator efficiency of 100%, the maximum electrical power is the turbine power.

$P_{\text{turbine}} = m_1(h_1 - h'_2) + (m_1 - m_2)(h'_2 - h'_3)$
$\qquad\qquad + (m_1 - m_2 - m_3)(h'_3 - h'_4)$

$\quad = \left(240{,}000 \dfrac{\text{lbm}}{\text{hr}}\right)\left(1368.3 \dfrac{\text{Btu}}{\text{lbm}} - 1150.1 \dfrac{\text{Btu}}{\text{lbm}}\right)$

$\quad + \left(240{,}000 \dfrac{\text{lbm}}{\text{hr}} - 72{,}000 \dfrac{\text{lbm}}{\text{hr}}\right)$

$\qquad \times \left(1150.1 \dfrac{\text{Btu}}{\text{lbm}} - 1108.6 \dfrac{\text{Btu}}{\text{lbm}}\right)$

$\quad + \left(240{,}000 \dfrac{\text{lbm}}{\text{hr}} - 72{,}000 \dfrac{\text{lbm}}{\text{hr}} - 24{,}000 \dfrac{\text{lbm}}{\text{hr}}\right)$

$\qquad \times \left(1108.6 \dfrac{\text{Btu}}{\text{lbm}} - 991.9 \dfrac{\text{Btu}}{\text{lbm}}\right)$

$\quad = 7.614 \times 10^7$ Btu/hr

Convert this to generator kW.

$$P_{\text{generator}} = \dfrac{7.614 \times 10^7 \dfrac{\text{Btu}}{\text{hr}}}{3412.9 \dfrac{\text{Btu}}{\text{hr-kW}}}$$

$$\qquad = 22.31 \text{ kW} \quad (22 \text{ kW})$$

The answer is C.

(c) All of the steam remaining after the bleeds passes through the condenser.

$\dot{m} = 240{,}000 \dfrac{\text{lbm}}{\text{hr}} - 72{,}000 \dfrac{\text{lbm}}{\text{hr}} - 24{,}000 \dfrac{\text{lbm}}{\text{hr}}$

$\quad = 144{,}000$ lbm/hr

From the steam tables, the specific volume of saturated water at 1 psia is

$v_5 = 0.01614$ ft³/lbm

$\dot{V} = \dfrac{\dot{m}}{\rho} = \dot{m}v$

$\quad = \dfrac{\left(144{,}000 \dfrac{\text{lbm}}{\text{hr}}\right)\left(0.01614 \dfrac{\text{ft}^3}{\text{lbm}}\right)}{\left(0.002228 \dfrac{\text{ft}^3\text{-min}}{\text{sec-gal}}\right)\left(3600 \dfrac{\text{sec}}{\text{hr}}\right)}$

$\quad = 289.8$ gal/min   (300 gal/min)

The answer is D.

(d) The heat transfer is the energy removed.

$q = \dot{m}(h'_4 - h_5)$

$\quad = \left(144{,}000 \dfrac{\text{lbm}}{\text{hr}}\right)\left(991.9 \dfrac{\text{Btu}}{\text{lbm}} - 69.7 \dfrac{\text{Btu}}{\text{lbm}}\right)$

$\quad = 1.328 \times 10^8$ Btu/hr   (130 MBtuh)

The answer is B.

(e) The pressure at the pump outlet is the same as the pressure in the steam generator. Although the connections are not shown, the boilerfeed pump pressurizes the entire mass flow rate, not just the condenser output.

$p_6 = p_1 = 300$ psia

$W = \dot{m}v_5\Delta p$

$\quad = \left(240{,}000 \dfrac{\text{lbm}}{\text{hr}}\right)\left(0.01614 \dfrac{\text{ft}^3}{\text{lbm}}\right)$

$\qquad \times \left(300 \dfrac{\text{lbf}}{\text{in}^2} - 1 \dfrac{\text{lbf}}{\text{in}^2}\right)\left(\dfrac{144 \dfrac{\text{in}^2}{\text{ft}^2}}{778 \dfrac{\text{ft-lbf}}{\text{Btu}}}\right)$

$\quad = 2.144$ Btu/hr   (210 kBtuh)

The answer is B.

(f) The energy content of the blowdown is

$q_7 = \dot{m}_7 h_7 = \left(48{,}000 \dfrac{\text{lbm}}{\text{hr}}\right)\left(1156.4 \dfrac{\text{Btu}}{\text{lbm}}\right)$

$\quad = 5.55 \times 10^7$ Btu/hr

The enthalpy of the water entering the steam generator is

$$h_6 = h_5 + (p_6 - p_5)v_5$$

$$= 69.7 \, \frac{\text{Btu}}{\text{lbm}} + \left(300 \, \frac{\text{lbf}}{\text{in}^2} - 1 \, \frac{\text{lbf}}{\text{in}^2}\right)$$

$$\times \left(0.01614 \, \frac{\text{ft}^3}{\text{lbm}}\right) \left(\frac{144 \, \frac{\text{in}^2}{\text{ft}^2}}{778 \, \frac{\text{ft-lbf}}{\text{Btu}}}\right)$$

$$= 70.6 \, \text{Btu/lbm}$$

The energy imparted to the motive steam is

$$q_{\text{steam}} = \dot{m}_{\text{steam}}(h_1 - h_6)$$

$$= \left(240{,}000 \, \frac{\text{Btu}}{\text{hr}}\right)\left(1368.3 \, \frac{\text{Btu}}{\text{lbm}} - 70.6 \, \frac{\text{Btu}}{\text{lbm}}\right)$$

$$= 3.114 \times 10^8 \, \text{Btu/hr}$$

The total ideal energy transfer in the steam generator is

$$q_{\text{steam generator}} = q_{\text{steam}} + q_7$$

$$= 3.114 \times 10^8 \, \frac{\text{Btu}}{\text{hr}} + 5.55 \times 10^7 \, \frac{\text{Btu}}{\text{hr}}$$

$$= 3.669 \times 10^8 \, \text{Btu/hr}$$

Considering the efficiencies of the furnace and steam generator, and the fact that 10% of the coal is noncombustible ash, the actual coal firing rate is

$$\dot{m}_{\text{coal}} = \frac{q_{\text{steam generator}}}{(\text{HHV})\eta_{\text{furnace}}\eta_{\text{steam generator}}}$$

$$= \frac{3.669 \times 10^8 \, \dfrac{\text{Btu}}{\text{hr}}}{\left(13{,}500 \, \dfrac{\text{Btu}}{\text{lbm}}\right)(0.60)(0.70)(1 - 0.10)}$$

$$= 7.190 \times 10^4 \, \text{lbm/hr} \quad (7.2 \times 10^4 \, \text{lbm/hr})$$

The answer is C.

(g) Assuming all of the oxygen is in the form of water, the ideal air/fuel ratio (based on the ash-free content) is

$$R_{\text{a/f,ideal}} = (34.5)\left(\frac{G_C}{3} + G_H - \frac{G_O}{8} + \frac{G_S}{8}\right)$$

$$= (34.5)\left(\frac{0.90}{3} + 0.04 - \frac{0.04}{8} + 0\right)$$

$$= 11.558 \, \text{lbm air/lbm fuel}$$

Using a table of reaction products, the volumes of the products of combustion per pound of fuel burned are

CO_2 from C:

$$\left(0.90 \, \frac{\text{lbm C}}{\text{lbm fuel}}\right)\left(31.63 \, \frac{\text{ft}^3 \, CO_2}{\text{lbm C}}\right)$$

$$= 28.47 \, \text{ft}^3 \, CO_2/\text{lbm fuel}$$

All hydrogen (free and bound) produces water vapor.

H_2O from H:

$$\left(0.04 \, \frac{\text{lbm H}}{\text{lbm fuel}}\right)\left(188.25 \, \frac{\text{ft}^3 \, H_2}{\text{lbm H}}\right)$$

$$= 7.53 \, \text{ft}^3 \, H_2O/\text{lbm fuel}$$

The oxygen in the fuel is in the form of water and does not contribute to free oxygen in the stack gases.

O_2 from 20% excess air:

$$V = \frac{mRT}{p}$$

$$= \frac{\begin{array}{c}(0.20)\left(11.558 \, \dfrac{\text{lbm air}}{\text{lbm fuel}}\right)\left(0.2315 \, \dfrac{\text{lbm } O_2}{\text{lbm air}}\right) \\[2mm] \times \left(48.29 \, \dfrac{\text{ft-lbf}}{\text{lbm-}°\text{R}}\right)(60°\text{F} + 460)\end{array}}{\left(14.7 \, \dfrac{\text{lbf}}{\text{in}^2}\right)\left(144 \, \dfrac{\text{in}^2}{\text{ft}^2}\right)}$$

$$= 6.35 \, \text{ft}^3 \, O_2/\text{lbm fuel}$$

N_2 from 120% theoretical air:

$$V = \frac{mRT}{p}$$

$$= \frac{\begin{array}{c}(1.20)\left(11.558 \, \dfrac{\text{lbm air}}{\text{lbm fuel}}\right)\left(0.7685 \, \dfrac{\text{lbm } N_2}{\text{lbm air}}\right) \\[2mm] \times \left(55.16 \, \dfrac{\text{ft-lbf}}{\text{lbm-}°\text{R}}\right)(60°\text{F} + 460)\end{array}}{\left(14.7 \, \dfrac{\text{lbf}}{\text{in}^2}\right)\left(144 \, \dfrac{\text{in}^2}{\text{ft}^2}\right)}$$

$$= 144.43 \, \text{ft}^3 \, N_2/\text{lbm fuel}$$

The total volume of stack gas per pound of fuel is

$$V_{\text{total}} = 28.47 \, \frac{\text{ft}^3 \, CO_2}{\text{lbm fuel}} + 7.53 \, \frac{\text{ft}^3 \, H_2O}{\text{lbm fuel}}$$

$$+ 6.35 \, \frac{\text{ft}^3 \, O_2}{\text{lbm fuel}} + 144.43 \, \frac{\text{ft}^3 \, N_2}{\text{lbm fuel}}$$

$$= 186.78 \, \text{ft}^3 \, \text{stack gas/lbm fuel}$$

The volumetric fraction of nitrogen in the stack gas is

$$B_{N^2} = \frac{144.43 \; \frac{\text{ft}^3 \; N_2}{\text{lbm fuel}}}{186.78 \; \frac{\text{ft}^3 \; \text{stack gas}}{\text{lbm fuel}}}$$

$$= 0.7733 \quad (77\%)$$

Note 1: This analysis can be based on the ash-free ultimate analysis. It is not necessary to convert to actual volumes per pound of fuel.

Note 2: The nitrogen in the ultimate analysis is problematic. Normally, this would indicate incombustible nitrates, rather than free nitrogen gas. However, the ultimate analysis was given on an ash-free basis. Arguments for including or not including the 2% nitrogen could be made. It is excluded in the solution to part (g), and included in part (h).

The answer is C.

(h) (Since (1) the stack gas is 80% nitrogen (by volume), (2) the specific heats of the other products essentially "balance out" each other, and (3) many other assumptions go into this calculation limiting accuracy of the result, good arguments could be made for using the specific heat of nitrogen and for skipping all of the calculations in this part.)

Using a table of reaction products, the masses of the products of combustion per pound of fuel burned are

CO_2 from C:

$$\left(0.90 \; \frac{\text{lbm C}}{\text{lbm fuel}}\right)\left(3.667 \; \frac{\text{lbm } CO_2}{\text{lbm C}}\right)$$

$$= 3.300 \; \text{lbm } CO_2/\text{lbm fuel}$$

All hydrogen (free and bound) contributes to water vapor in the stack.

H_2O from H:

$$\left(0.04 \; \frac{\text{lbm H}}{\text{lbm fuel}}\right)\left(8.936 \; \frac{\text{lbm } H_2}{\text{lbm H}}\right)$$

$$= 0.357 \; \text{lbm } H_2/\text{lbm fuel}$$

N_2 from C, with 20% excess:

$$(1.2)\left(0.90 \; \frac{\text{lbm C}}{\text{lbm fuel}}\right)\left(8.883 \; \frac{\text{lbm } N_2}{\text{lbm C}}\right)$$

$$= 9.594 \; \text{lbm } N_2/\text{lbm fuel}$$

N_2 from H_2, with 20% excess:

Only the free hydrogen contributes to nitrogen in the stack gas. Hydrogen in the form of water does not burn.

$$\% \; \text{free } H_2 = G_{H_2} - \frac{G_{O_2}}{8} = 0.04 - \frac{0.04}{8} = 0.035$$

$$(1.2)\left(0.035 \; \frac{\text{lbm } H_2}{\text{lbm fuel}}\right)\left(26.29 \; \frac{\text{lbm } N_2}{\text{lbm } H_2}\right)$$

$$= 1.104 \; \text{lbm } N_2/\text{lbm fuel}$$

N_2 from the fuel:

$$0.02 \; \text{lbm } N_2/\text{lbm fuel}$$

O_2 from excess air:

$$(0.20)\left(11.558 \; \frac{\text{lbm air}}{\text{lbm fuel}}\right)(0.2315)$$

$$= 0.535 \; \text{lbm } O_2/\text{lbm fuel}$$

Summarizing the products of combustion per ash-free pound of fuel:

N_2: $9.594 \; \frac{\text{lbm } N_2}{\text{lbm fuel}} + 1.104 \; \frac{\text{lbm } N_2}{\text{lbm fuel}} + 0.02 \; \frac{\text{lbm } N_2}{\text{lbm fuel}}$

$$= 10.718 \; \text{lbm } N_2/\text{lbm fuel}$$

CO_2: $3.300 \; \text{lbm } CO_2/\text{lbm fuel}$

H_2O: $0.357 \; \text{lbm } H_2/\text{lbm fuel}$

O_2: $0.535 \; \text{lbm } O_2/\text{lbm fuel}$

total: $14.910 \; \text{lbm stack gases/lbm fuel}$

Since the stack gases are stated to be ideal gases, their specific heats do not vary with temperature. (Otherwise, an estimate of the combustion temperature would have to be made prior to this calculation.) The mean specific heat is

$$c_{p,\text{mean}} = \frac{\begin{array}{c}(10.718)(0.249) + (3.300)(0.207) \\ + (0.357)(0.445) + (0.535)(0.220)\end{array}}{14.910}$$

$$= 0.243 \; \text{Btu/lbm-}°\text{F}$$

In the United States, it is traditional to report the higher heating value of the coal. However, the maximum theoretical flame temperature should be based on the lower heating value, since the water vapor does not condense within the combustion area. Since the lower and higher heating values of coal do not usually differ by more than 5%, and since great accuracy is not called for in this calculation, use 95% of the higher heating value.

The coal is 10% noncombustible ash, so only 90% of it contributes to producing stack gases.

$$T_{\max} = T_i + \frac{\text{heat of combustion}}{m_{\text{products}}c_{p,\text{mean}}}$$

$$= 60°\text{F} + \frac{(0.95)\left(13{,}500\,\dfrac{\text{Btu}}{\text{lbm}}\right)}{(0.90)\left(14.910\,\dfrac{\text{lbm stack gases}}{\text{lbm fuel}}\right)}$$
$$\times \left(0.243\,\dfrac{\text{Btu}}{\text{lbm-°F}}\right)$$

$$= 3993°\text{F} \quad (4000°\text{F})$$

This assumes that all of the combustion energy goes into the stack gases.

The answer is C.

(i) From part (h), the mass of water vapor per pound of ash-free fuel is 0.357 lbm H_2/lbm fuel. Since the coal is 10% noncombustible ash, the mass of water vapor per pound of coal is

$$\frac{0.357\,\dfrac{\text{lbm }H_2}{\text{lbm fuel}}}{0.90\,\dfrac{\text{lbm fuel}}{\text{lbm coal}}} = 0.397 \quad (0.40)$$

The answer is B.

(j) From Dalton's law, partial pressure is mole weighted. The mole and volumetric fractions are the same. The volumetric fraction of the water vapor in the stack gas is

$$x_{H_2O} = \frac{7.53\,\dfrac{\text{ft}^3\ H_2O}{\text{lbm fuel}}}{186.78\,\dfrac{\text{ft}^3\text{ stack gas}}{\text{lbm fuel}}} = 0.040$$

The partial pressure of the water vapor is

$$p_{H_2O} = x_{H_2O}p_{\text{total}} = (0.040)\left(15.0\,\dfrac{\text{lbf}}{\text{in}^2}\right)$$
$$= 0.6\ \text{lbf/in}^2$$

From the steam tables, the saturation temperature corresponding to this pressure is 85.19°F (85°F).

The answer is A.

14 SOLUTION

This corresponds to a standard vapor compression cycle except for the superheat.

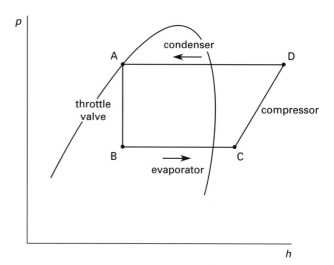

(a) Use either a table or graph of R-134a properties. The properties at the inlet to the compressor are

$$p_C = 50\ \text{kPa} \quad (0.05\ \text{MPa};\ 0.5\ \text{bar})$$
$$T_C = 232.7°\text{K} - 273 + 20 = -20.3°\text{C} \quad (-20°\text{C})$$

Read h_C from the graph, from a table of superheat values, or as $h_g + c_p(20°\text{C})$.

$$h_C = 374.1\,\frac{\text{kJ}}{\text{kg}} + \left(0.748\,\frac{\text{kJ}}{\text{kg·K}}\right)(20°\text{C})$$
$$= 389\ \text{kJ/kg} \quad (390\ \text{kJ/kg})$$

The answer is B.

(b) Use the graph. From point C, follow a line of constant entropy up to 600 kPa.

$$p_D = 600\ \text{kPa} \quad (0.6\ \text{MPa};\ 6\ \text{bar})$$
$$T_D = 57°\text{C}$$
$$h_D = 445\ \text{kJ/kg}$$

With the compressor inefficiency, the enthalpy entering the condenser is

$$h_D' = h_C + \frac{h_D - h_C}{\eta_{s,\text{compressor}}}$$

$$= 389\,\frac{\text{kJ}}{\text{kg}} + \frac{445\,\dfrac{\text{kJ}}{\text{kg}} - 389\,\dfrac{\text{kJ}}{\text{kg}}}{0.80}$$

$$= 459\ \text{kJ/kg} \quad (460\ \text{kJ/kg})$$

The answer is D.

(c) From tables or the graph,

$$p_A = 600 \text{ kPa} \quad (0.6 \text{ MPa; } 6 \text{ bar})$$
$$T_A = 294.7 \text{K}$$
$$h_A = 229.7 \text{ kJ/kg} \quad [\text{saturated liquid}]$$
$$T_A = 294.7 \text{K} - 273 = 21.7°\text{C} \quad (22°\text{C})$$

The answer is C.

(d) Since the refrigerant experiences a throttling process, the theoretical enthalpy at point B is the same as at point A.

$$p_B = p_C = 50 \text{ kPa} \quad (0.05 \text{ MPa; } 0.5 \text{ bar})$$
$$h_B = h_A = 229.7 \text{ kJ/kg}$$
$$T_B = 232.7 \text{K}$$

The refrigeration effect occurs between points B and C. The refrigerant flow is

$$\dot{m}_{\text{refrigerant}} = \frac{\dot{q}}{h_C - h_B}$$

$$= \frac{(6 \text{ tons})\left(12{,}000 \dfrac{\text{Btu}}{\text{h-ton}}\right)\left(0.2931 \dfrac{\text{W-h}}{\text{Btu}}\right)}{\left(389 \dfrac{\text{kJ}}{\text{kg}} - 229.7 \dfrac{\text{kJ}}{\text{kg}}\right)\left(1000 \dfrac{\text{J}}{\text{kJ}}\right)}$$

$$= 0.1325 \text{ kg/s} \quad (0.13 \text{ kg/s})$$

The answer is A.

(e) External power is required between points C and D.

$$P = \frac{\dot{m}_{\text{refrigerant}}(h'_D - h_C)}{\eta_{m,\text{compressor}}}$$

$$= \frac{\left(0.20 \dfrac{\text{kg}}{\text{s}}\right)\left(459 \dfrac{\text{kJ}}{\text{kg}} - 389 \dfrac{\text{kJ}}{\text{kg}}\right)}{0.90}$$

$$= 15.56 \text{ kW} \quad (16 \text{ kW})$$

The answer is C.

(f) The energy balance on the condenser is

$$\dot{m}_{\text{water}} c_p (T_{\text{out}} - T_{\text{in}}) = \dot{m}_{\text{refrigerant}}(h'_D - h_A)$$

$$\dot{m}_{\text{water}} \left(4.19 \dfrac{\text{kJ}}{\text{kg·K}}\right)(22°\text{C} - 16°\text{C})$$

$$= \left(0.20 \dfrac{\text{kg}}{\text{s}}\right)\left(459 \dfrac{\text{kJ}}{\text{kg}} - 229.7 \dfrac{\text{kJ}}{\text{kg}}\right)$$

$$\dot{m}_{\text{water}} = 1.82 \text{ kg/s}$$

$$\dot{V} = \frac{\dot{m}}{\rho} = \frac{\left(1.82 \dfrac{\text{kg}}{\text{s}}\right)\left(1000 \dfrac{\text{L}}{\text{m}^3}\right)}{998 \dfrac{\text{kg}}{\text{m}^3}}$$

$$= 1.82 \text{ L/s} \quad (2 \text{ L/s})$$

The answer is C.

(g) $$c_p = \frac{\Delta h}{\Delta T} = \frac{h'_D - h_C}{T_D - T_C}$$

$$= \frac{459 \dfrac{\text{kJ}}{\text{kg}} - 389 \dfrac{\text{kJ}}{\text{kg}}}{57°\text{C} - (-20.3°\text{C})}$$

$$= 0.9056 \text{ kJ/kg·K} \quad (0.91 \text{ kJ/kg·K})$$

The answer is D.

(h) The specific volume entering the compressor is

$$v_C = 0.4 \text{ m}^3/\text{kg}$$

$$\dot{V} = \dot{m}_{\text{refrigerant}} v_C = \left(0.20 \dfrac{\text{kg}}{\text{s}}\right)\left(0.4 \dfrac{\text{m}^3}{\text{kg}}\right)$$

$$= 0.08 \text{ m}^3/\text{s} \quad (80 \text{ L/s})$$

The answer is A.

(i) With a single-acting compressor, there is one compression stroke every two revolutions. So, the frequency of compression strokes is

$$N = \frac{250 \dfrac{\text{rev}}{\text{min}}}{\left(2 \dfrac{\text{rev}}{\text{compression stroke}}\right)\left(60 \dfrac{\text{s}}{\text{min}}\right)}$$

$$= 2.083 \text{ strokes/s}$$

Since the stroke length (L) is twice the bore diameter, the volumetric piston displacement per power stroke is

$$V = AL = \left(\frac{\pi}{4}\right)D^2 L = \left(\frac{\pi}{4}\right)D^2(2D) = \left(\frac{\pi}{2}\right)D^3$$

Assuming a volumetric efficiency of 100%, the ideal volumetric flow rate is NV.

$$\dot{V} = NV = N\left(\frac{\pi}{2}\right)D^3$$

$$0.08 \dfrac{\text{m}^3}{\text{s}} = \left(2.083 \dfrac{\text{strokes}}{\text{s}}\right)\left(\frac{\pi}{2}\right)D^3$$

$$D = 0.290 \text{ m} \quad (290 \text{ mm})$$

(This is probably too large to be a practical design.)

The answer is D.

(j)
$$\eta_{\text{volumetric}} = \frac{V_{\text{ideal}}}{V_{\text{actual}}} = \frac{N\left(\frac{\pi}{2}\right)D_{\text{ideal}}^3}{N\left(\frac{\pi}{2}\right)D_{\text{actual}}^3}$$

$$= \frac{D_{\text{ideal}}^3}{D_{\text{actual}}^3} = \left(\frac{290 \text{ mm}}{310 \text{ mm}}\right)^3$$

$$= 0.819 \quad (82\%)$$

The answer is C.

15 SOLUTION

(a) The saturation temperature of steam at 1 atm is 212°F. The thermal conductivity of carbon steel pipe at 212°F is approximately 29.2 Btu/hr-ft-°F. Lower values (e.g., 25–26 Btu/hr-ft-°F) would apply only to alloy steels.

The answer is C.

(b) The dimensions of $1\frac{1}{2}$ in, 16 gage BWG pipe are

$$d_i = 1.37 \text{ in}$$
$$t = 0.065 \text{ in}$$
$$d_o = d_i + 2t = 1.37 \text{ in} + (2)(0.065 \text{ in}) = 1.5 \text{ in}$$
$$r_i = \frac{d_i}{2} = \frac{1.37 \text{ in}}{2} = 0.685 \text{ in}$$
$$r_o = \frac{d_o}{2} = \frac{1.50 \text{ in}}{2} = 0.75 \text{ in}$$

The thermal resistance of a curved layer is

$$R_{\text{th}} = \frac{\ln\left(\frac{r_o}{r_i}\right)}{2\pi k L} = \frac{\ln\left(\frac{0.75 \text{ in}}{0.685 \text{ in}}\right)}{(2\pi)\left(29 \dfrac{\text{Btu}}{\text{hr-ft-°F}}\right)(1 \text{ ft})}$$

$$= 4.98 \times 10^{-4} \text{ hr-°F/Btu} \quad (0.0005 \text{ hr-°F/Btu})$$

The answer is B.

(c) The overall heat transfer coefficient is

$$\frac{1}{U_o} = \frac{1}{h_o} + \left(\frac{r_o}{k}\right)\ln\left(\frac{r_o}{r_i}\right) + \frac{r_o}{r_i h_i}$$

$$= \frac{1}{2.0 \dfrac{\text{Btu}}{\text{hr-ft}^2\text{-°F}}}$$

$$+ \left(\frac{0.75 \text{ in}}{\left(12 \dfrac{\text{in}}{\text{ft}}\right)\left(29 \dfrac{\text{Btu}}{\text{hr-ft-°F}}\right)}\right)\ln\left(\frac{0.75 \text{ in}}{0.685 \text{ in}}\right)$$

$$+ \frac{0.75 \text{ in}}{(0.685 \text{ in})\left(1500 \dfrac{\text{Btu}}{\text{hr-ft}^2\text{-°F}}\right)}$$

$$= 0.501 \text{ hr-ft}^2\text{-°F/Btu}$$

$$U_o = \frac{1}{0.501 \dfrac{\text{hr-ft}^2\text{-°F}}{\text{Btu}}}$$

$$= 2.0 \text{ Btu/hr-ft}^2\text{-°F}$$

The answer is B.

(d) The heat loss over the entire length of pipe is

$$Q = U_o A_o \Delta T = U_o \pi D L \Delta T$$

$$= \left(2.0 \dfrac{\text{Btu}}{\text{hr-ft}^2\text{-°F}}\right)\pi\left(\frac{1.5 \text{ in}}{12 \dfrac{\text{in}}{\text{ft}}}\right)$$

$$\times (120 \text{ ft})(212\text{°F} - 60\text{°F})$$

$$= 14{,}326 \text{ Btu/hr}$$

The volume of the steam in the pipe is

$$V = A_i L = \left(\frac{\pi}{4}\right)\left(\frac{1.37 \text{ in}}{12 \dfrac{\text{in}}{\text{ft}}}\right)^2 (120 \text{ ft})$$

$$= 1.228 \text{ ft}^3$$

The heat loss for 45 min (0.75 hr) per unit volume is

$$\frac{(0.75 \text{ hr})\left(14{,}326 \dfrac{\text{Btu}}{\text{hr}}\right)}{1.228 \text{ ft}^3} = 8750 \text{ Btu/ft}^3$$
$$(8800 \text{ Btu/ft}^3)$$

The answer is A.

(e) The specific volume of saturated 212°F steam is

$$v = 26.80 \text{ ft}^3/\text{lbm}$$

The mass of the steam in the pipe is

$$m = \rho V = \frac{V}{v} = \frac{1.228 \text{ ft}^3}{26.80 \dfrac{\text{ft}^3}{\text{lbm}}}$$

$$= 0.0458 \text{ lbm}$$

The initial enthalpy of the saturated 212°F steam is 970.3 Btu/lbm.

Since the steam is initially saturated, any cooling will result in some condensation. At 60°F, all of the steam will have condensed and there will be 0.0458 lbm of 60°F liquid water in the pipe.

The enthalpy of 60°F water is approximately 28.08 Btu/lbm.

The total energy loss is

$$(0.0458 \text{ lbm}) \left(970.3 \ \frac{\text{Btu}}{\text{lbm}} - 28.08 \ \frac{\text{Btu}}{\text{lbm}}\right) = 43.15 \text{ Btu}$$

The time to cool is

$$t = \frac{(43.15 \text{ Btu}) \left(60 \ \frac{\text{min}}{\text{hr}}\right)}{(0.27) \left(14{,}326 \ \frac{\text{Btu}}{\text{hr}}\right)}$$

$$= 0.67 \text{ min} \quad (1 \text{ min})$$

The answer is A.

(f) Disregard the insignificant thermal resistance of the steel pipe. Assuming a perfect circular wrap, the dimensions of the insulation are

$$r_{i,\text{insulation}} = r_{o,\text{pipe}} = 0.75 \text{ in}$$

$$r_{o,\text{insulation}} = r_{i,\text{insulation}} + t_{\text{insulation}}$$

$$= 0.75 \text{ in} + 2 \text{ in}$$

$$= 2.75 \text{ in}$$

Assume the temperature of the insulation is approximately 70°F. Then, the average temperature of the insulation will be

$$T_{\text{mean,insulation}} = \frac{212°\text{F} + 70°\text{F}}{2}$$

$$= 141°\text{F} \quad (150°\text{F})$$

At this temperature, the thermal conductivity of 85% magnesia insulation is approximately 0.032 Btu/hr-ft-°F.

The thermal resistance of a curved layer is

$$R_{\text{th}} = \frac{\ln\left(\frac{r_o}{r_i}\right)}{2\pi k L} = \frac{\ln\left(\frac{2.75 \text{ in}}{0.75 \text{ in}}\right)}{(2\pi)\left(0.032 \ \frac{\text{Btu}}{\text{hr-ft-°F}}\right)(1 \text{ ft})}$$

$$= 6.46 \text{ hr-°F/Btu} \quad (6.5 \text{ °F-sec/Btu})$$

The answer is D.

(g) The internal film coefficient is "large," so it will not contribute much to the thermal resistance. The insulation and outside film coefficient will have the largest effects on the thermal resistance. Since the surface temperature is close to the ambient temperature, laminar convection is the most likely. Prove that this is the case.

The outside film will be at a temperature of

$$T_{\text{film}} = \tfrac{1}{2}(T_s + T_\infty) = \left(\tfrac{1}{2}\right)(70°\text{F} + 60°\text{F}) = 65°\text{F}$$

From tables for 65°F air, approximately,

$$\frac{g\beta\rho^2}{\mu^2} = 2.46 \times 10^6 \ \frac{1}{\text{ft}^3\text{-°F}}$$

$$\text{Pr} = 0.72$$

The characteristic length is the outside diameter of the insulation.

$$D_o = 2r_o = \frac{(2)(2.75 \text{ in})}{12 \ \frac{\text{in}}{\text{ft}}} = 0.4583 \text{ ft}$$

The Grashof number is

$$\text{Gr} = \frac{L^3 g\beta\rho^2 (T_s - T_\infty)}{\mu^2}$$

$$= (0.4583 \text{ ft})^3 \left(2.46 \times 10^6 \ \frac{1}{\text{ft}^3\text{-°F}}\right)(70°\text{F} - 60°\text{F})$$

$$= 2.37 \times 10^6$$

The product of the Grashof and Prandtl numbers is

$$\text{GrPr} = (2.37 \times 10^6)(0.72) = 1.71 \times 10^6$$

This is within the range for natural convection. A correlation for the natural convective outside film coefficient is

$$h_o = (0.27)\left(\frac{T_s - T_\infty}{d_o}\right)^{\frac{1}{4}} = (0.27)\left(\frac{70°\text{F} - 60°\text{F}}{0.4583 \text{ ft}}\right)^{\frac{1}{4}}$$

$$= 0.584 \text{ Btu/hr-ft}^2\text{-°F}$$

Disregarding the insignificant thermal resistance of the pipe material, the overall heat transfer coefficient is

$$\frac{1}{U_o} = \frac{1}{h_o} + \left(\frac{r_{o,\text{ins}}}{k_{\text{ins}}}\right) \ln\left(\frac{r_{o,\text{ins}}}{r_{i,\text{ins}}}\right) + \frac{r_o}{r_{i,\text{pipe}} h_i}$$

$$= \frac{1}{0.584 \ \frac{\text{Btu}}{\text{hr-ft}^2\text{-°F}}}$$

$$+ \left(\frac{2.75 \text{ in}}{\left(12 \ \frac{\text{in}}{\text{ft}}\right)\left(0.032 \ \frac{\text{Btu}}{\text{hr-ft-°F}}\right)}\right) \ln\left(\frac{2.75 \text{ in}}{0.75 \text{ in}}\right)$$

$$+ \frac{2.75 \text{ in}}{(0.685 \text{ in})\left(1500 \ \frac{\text{Btu}}{\text{hr-ft}^2\text{-°F}}\right)}$$

$$= 11.02 \text{ hr-ft}^2\text{-°F/Btu}$$

$$U_o = \frac{1}{11.02 \ \frac{\text{hr-ft}^2\text{-°F}}{\text{Btu}}}$$

$$= 0.0907 \text{ Btu/hr-ft}^2\text{-°F}$$

The heat loss from the insulated pipe is

$$Q = U_o A_o (T_i - T_\infty)$$

$$= \left(0.0907 \, \frac{\text{Btu}}{\text{hr-ft}^2\text{-}°\text{F}} \right) (2\pi) \left(\frac{2.75 \text{ in}}{12 \, \frac{\text{in}}{\text{ft}}} \right)$$

$$\times (120 \text{ ft})(212°\text{F} - 60°\text{F})$$

$$= 2382 \text{ Btu/hr} \quad (2400 \text{ Btuh})$$

The answer is C.

(h) The traditional correlation for fluid flow over a single tube is of the form

$$\text{Nu} = C_1 (\text{Re}_d)^n \text{Pr}^{\frac{1}{3}}$$

This corresponds to Eq. II.

Since the $\text{Pr}^{\frac{1}{3}}$ is close to 1, the last term can be omitted.

$$\text{Nu} = C_1 (\text{Re}_d)^n$$

This corresponds to Eq. I.

Equation III is the Nusselt equation for turbulent flow through circular pipes.

The answer is C.

(i) As in part (g), the film is evaluated at 65°F.

For 65°F air,

$$\text{Pr} = 0.72$$

$$\nu = 0.163 \times 10^{-3} \text{ ft}^2/\text{sec}$$

$$k = 0.0147 \text{ Btu/hr-ft-}°\text{F}$$

The Reynolds number based on the outside diameter of the bare pipe is

$$\text{Re} = \frac{D\text{v}}{\nu} = \frac{\left(\frac{1.50 \text{ in}}{12 \, \frac{\text{in}}{\text{ft}}} \right) \left(30 \, \frac{\text{ft}}{\text{sec}} \right)}{0.163 \times 10^{-3} \, \frac{\text{ft}^2}{\text{sec}}} = 2.30 \times 10^4$$

For this range, the corresponding correlation constants are

$$C_1 = 0.193$$

$$n = 0.618$$

$$\text{Nu} = \frac{hD}{k} = C_1 (\text{Re}_d)^n \text{Pr}^{\frac{1}{3}}$$

$$h = \left(\frac{k}{d} \right) C_1 (\text{Re}_d)^n \text{Pr}^{\frac{1}{3}}$$

$$= \left(\frac{0.0147 \, \frac{\text{Btu}}{\text{hr-ft-}°\text{F}}}{\frac{1.5 \text{ in}}{12 \, \frac{\text{in}}{\text{ft}}}} \right) (0.193)(23,000)^{0.618}(0.72)^{\frac{1}{3}}$$

$$= 10.09 \text{ Btu/hr-ft}^2\text{-}°\text{F} \quad (10 \text{ Btu/hr-ft}^2\text{-}°\text{F})$$

The answer is B.

(j) Use the correlation from part (i).

$$h = \left(\frac{k}{d} \right) C_1 (\text{Re}_d)^n \text{Pr}^{\frac{1}{3}}$$

The film coefficient is related to the 0.618th power of velocity.

$$\frac{h_2}{h_1} = \left(\frac{\text{v}_2}{\text{v}_1} \right)^{0.618} = (2)^{0.618}$$

$$= 1.53 \quad (50\%) \quad [\text{a } 53\% \text{ increase}]$$

The answer is B.

16 SOLUTION

(a) Locate the outside air at 90°F and 45% relative humidity on the psychrometric chart. Read:

$$T_{\text{wb}} = 72.7°\text{F} \quad (73°\text{F})$$

$$\omega = 0.0134 \text{ lbm/lbm} \quad (94 \text{ gr/lbm})$$

$$v = 14.13 \text{ ft}^3/\text{lbm}$$

$$T_{\text{dp}} = 65°\text{F}$$

The answer is C.

(b) Use data from part (a).

$$\dot{m}_w = \omega m_a = \omega \dot{V} \rho = \frac{\omega \dot{V}}{v}$$

$$= \frac{\left(0.0134 \, \frac{\text{lbm}}{\text{lbm}} \right) \left(3000 \, \frac{\text{ft}^3}{\text{min}} \right) \left(60 \, \frac{\text{min}}{\text{hr}} \right)}{14.13 \, \frac{\text{ft}^3}{\text{lbm}}}$$

$$= 170.7 \text{ lbm/hr} \quad (170 \text{ lbm/hr})$$

The answer is D.

(c) From part (a), $T_{\text{dp}} = 65°\text{F}$.

The answer is C.

(d) Locate the starting points (80°F dry bulb and 70% relative humidity) and ending points (56°F dry bulb and 90% relative humidity). Read:

$$\omega_{in} = 0.0154 \text{ lbm/lbm}$$
$$v_{in} = 13.94 \text{ ft}^3/\text{lbm}$$
$$T_{dp,in} = 69.4°\text{F}$$
$$\omega_{out} = 0.00857 \text{ lbm/lbm}$$
$$T_{dp,out} = 52.9°\text{F}$$

Draw a straight line between the two points. Translate the line through the chart's pivot point and read the sensible heat factor as 0.44 from the scale. (The SHF reads slightly different on different charts. It is approximately 0.44 on a Carrier chart; it is approximately 0.46 on a Trane chart.)

The answer is C.

(e) The mass removed is

$$\dot{m}_w = \dot{m}_{a,in}(\omega_{in} - \omega_{out}) = \left(\frac{\dot{V}_{a,in}}{v_{in}}\right)(\omega_{in} - \omega_{out})$$

$$= \left(\frac{\left(15{,}000 \dfrac{\text{ft}^3}{\text{min}}\right)\left(60 \dfrac{\text{min}}{\text{hr}}\right)}{13.94 \dfrac{\text{ft}^3}{\text{lbm}}}\right)$$

$$\times \left(0.0154 \frac{\text{lbm}}{\text{lbm}} - 0.00857 \frac{\text{lbm}}{\text{lbm}}\right)$$

$$= 441 \text{ lbm/hr} \quad (440 \text{ lbm/hr})$$

The answer is C.

(f) The dew point of the conditioner output is 52.9°F. The mass of the conditioned air does not change through the conditioner.

$$m_1 = \frac{\dot{V}}{v} = \frac{15{,}000 \dfrac{\text{ft}^3}{\text{min}}}{13.94 \dfrac{\text{ft}^3}{\text{lbm}}} = 1076 \text{ lbm/min}$$

The dew point of the outside air is 65°F. The mass of the outside air is

$$m_2 = \frac{\dot{V}}{v} = \frac{3000 \dfrac{\text{ft}^3}{\text{min}}}{14.13 \dfrac{\text{ft}^3}{\text{lbm}}} = 212.3 \text{ lbm/min}$$

Use the lever rule.

$$T_{dp,mixture} = T_{dp,1} + \left(\frac{m_2}{m_1 + m_2}\right)(T_{dp,2} - T_{dp,1})$$

$$= 52.9°\text{F} + \left(\frac{212.3 \text{ lbm}}{212.3 \text{ lbm} + 1076 \text{ lbm}}\right)$$

$$\times (65°\text{F} - 52.9°\text{F})$$

$$= 52.9°\text{F} + (0.1648)(65°\text{F} - 52.9°\text{F})$$

$$= 54.9°\text{F} \quad (55°\text{F})$$

The answer is B.

(g) Various methods can be used to find the remaining properties from the psychrometric chart. However, the lever rule can also be used.

$$\omega_{mixture} = \omega_1 + \left(\frac{m_2}{m_1 + m_2}\right)(\omega_2 - \omega_1)$$

$$= 0.00857 \frac{\text{lbm}}{\text{lbm}} + (0.1648)$$

$$\times \left(0.0134 \frac{\text{lbm}}{\text{lbm}} - 0.00857 \frac{\text{lbm}}{\text{lbm}}\right)$$

$$= 0.00937 \text{ lbm/lbm} \quad (0.0094 \text{ lbm/lbm})$$

The answer is D.

(h) $$T_{db,mixture} = T_{db,1} + \left(\frac{m_2}{m_1 + m_2}\right)(T_{db,2} - T_{db,1})$$
$$= 56°\text{F} + (0.1648)(90°\text{F} - 56°\text{F})$$
$$= 61.6°\text{F} \quad (62°\text{F})$$

The answer is A.

(i) The total air entering the conditioned space is

$$\dot{V} = 15{,}000 \frac{\text{ft}^3}{\text{min}} + 3000 \frac{\text{ft}^3}{\text{min}} = 18{,}000 \text{ ft}^3/\text{min}$$

Using appropriate approximations, the "traditional" sensible heating-ventilation relationship is

$$\dot{V}_{cfm} = \frac{\dot{q}_{s,Btu/hr}}{\left(1.08 \dfrac{\text{Btu-min}}{\text{ft}^3\text{-hr-}°\text{F}}\right)(T_{out} - T_{in})}$$

$$18{,}000 \frac{\text{ft}^3}{\text{min}} = \frac{2 \times 10^5 \dfrac{\text{Btu}}{\text{hr}}}{\left(1.08 \dfrac{\text{Btu-min}}{\text{ft}^3\text{-hr-}°\text{F}}\right)(T_{out} - 61.6°\text{F})}$$

$$T_{out} = 71.9°\text{F} \quad (72°\text{F})$$

The answer is B.

(j) This requires locating the points (61.6°F db and 54.9°F dp, and 75°F db and 70% rh) on the psychrometric chart. The sensible heat ratio for this process is found from the chart in the same manner as for part (d) as 0.36. This is the fraction of the load that is sensible. The fraction of the load that is latent is $1 - 0.36 = 0.64$.

The answer is D.

17 SOLUTION

(Slightly different answers are obtained from different tables of HVAC data. This solution relies on data from the ASHRAE *Handbook of Fundamentals*.)

(a) The crack length in a double-hung window is the perimeter of the window plus the horizontal break between the upper and lower panes.

$$L = 3 \text{ ft} + 3 \text{ ft} + 5 \text{ ft} + 5 \text{ ft} + 3 \text{ ft} = 19 \text{ ft}$$

The answer is C.

(b) The crack length is given as 16 ft for the casement window. From a suitable table, the crack length coefficient for a good quality wood casement window at this wind speed (pressure) is approximately 0.50 ft³/min-ft. The infiltration rate per window is

$$\dot{V} = BL = \left(0.50 \; \frac{\text{ft}^3}{\text{min-ft}}\right)(16 \text{ ft})\left(60 \; \frac{\text{min}}{\text{hr}}\right)$$

$$= 480 \text{ ft}^3/\text{hr} \quad (500 \text{ ft}^3/\text{hr})$$

(Note: This assumes a modern energy-efficient casement window. The crack length coefficient may be reported as higher in some older HVAC books. However, this would not change the answer choice.)

The answer is D.

(c) Infiltration from double-hung windows in a caulked-masonry installation comes from two sources: from around the sash and from around the window caulked frame-masonry joint. Each of these sources has its own crack length coefficient. From a suitable table, the crack length coefficients are

sash crack: 28 ft³/hr-ft
frame wall crack: 3 ft³/hr-ft

The difference in infiltrations is

$$\Delta\dot{V} = \dot{V}_{\text{double hung}} - \dot{V}_{\text{casement}} = BL - 430 \; \frac{\text{ft}^3}{\text{hr}}$$

$$= \left(28 \; \frac{\text{ft}^3}{\text{hr-ft}}\right)(19 \text{ ft}) + \left(3 \; \frac{\text{ft}^3}{\text{hr-ft}}\right)(16 \text{ ft}) - 430 \; \frac{\text{ft}^3}{\text{hr}}$$

$$= 150 \text{ ft}^3/\text{hr}$$

The answer is C.

(d) The enthalpy of the outside air cannot be determined because the moisture content is unknown. Use an approach based on an average specific heat and dry air.

$$Q = \dot{m}c_p\Delta T = \dot{V}\rho c_p\Delta T = \dot{V}\left(1.08 \; \frac{\text{Btu-min}}{\text{ft}^3\text{-hr-}^\circ\text{F}}\right)\Delta T$$

$$= \frac{\left(120 \; \dfrac{\text{ft}^3}{\text{hr}}\right)\left(1.08 \; \dfrac{\text{Btu-min}}{\text{ft}^3\text{-hr-}^\circ\text{F}}\right)(89^\circ\text{F} - 80^\circ\text{F})}{60 \; \dfrac{\text{min}}{\text{hr}}}$$

$$= 19.4 \text{ Btu/hr} \quad (20 \text{ Btu/hr})$$

(Notice that this result is for dry air only, and that the heat content due to moisture could be substantial.)

The answer is A.

(e) Use a psychrometric chart. Locate the intersection of 80°F dry bulb and 50% relative humidity.

$$\omega_{id} = 0.0112 \text{ lbm/lbm} \quad (0.011 \text{ lbm/lbm})$$

The answer is B.

(f) Use a psychrometric chart. Locate the intersection of 89°F dry bulb and 75°F wet bulb.

$$\omega_{od} = 0.0155 \text{ lbm/lbm}$$

Use the value of ω_{id} from part (e).

$$\dot{q}_{l,\text{Btu/hr}} = \dot{V}_{\text{cfm}}\left(4775 \; \frac{\text{Btu-min}}{\text{ft}^3\text{-hr}}\right)(\omega_{od} - \omega_{id})$$

$$= \frac{\left(120 \; \dfrac{\text{ft}^3}{\text{hr}}\right)\left(4775 \; \dfrac{\text{Btu-min}}{\text{ft}^3\text{-hr}}\right)}{60 \; \dfrac{\text{min}}{\text{hr}}}$$
$$\times \left(0.0155 \; \dfrac{\text{lbm}}{\text{lbm}} - 0.0112 \; \dfrac{\text{lbm}}{\text{lbm}}\right)$$

$$= 41.07 \text{ Btu/hr} \quad (40 \text{ Btu/hr})$$

The answer is B.

(g) Use the equivalent full-load hour method. The annual savings per window are

$$
\text{savings} = \frac{\left(40 \; \dfrac{\text{Btu}}{\text{hr-window}} \right) (10 \; \text{windows})}{\left(8 \; \dfrac{\text{Btu}}{\text{W-hr}} \right) \left(1000 \; \dfrac{\text{W}}{\text{kW}} \right)}
$$

$$
= \$7.50/\text{yr}
$$

For 10 windows per condominium, the annual savings would be

$$
(10 \; \text{windows}) \left(\frac{\$0.75}{\text{yr-window}} \right) = \$7.50/\text{yr} \quad (\$8 \; \text{per year})
$$

The answer is A.

(h) To encourage users (typically in large buildings) not to place demands on the electrical grid, utilities may levy demand charges. Demand charges are in addition to the charge for electrical power used. Demand charges are based on the peak load during the heating season.

$$
\text{savings} = \frac{\left(40 \; \dfrac{\text{Btu}}{\text{hr-window}} \right) (10 \; \text{windows})}{\left(8 \; \dfrac{\text{Btu}}{\text{W-hr}} \right) \left(1000 \; \dfrac{\text{W}}{\text{kW}} \right)}
$$

$$
= \$3.00/\text{yr}
$$

The answer is A.

(i) The sensible heat loss reduction is

$$
\dot{q}_{s,\text{Btu/hr}} = \dot{V}_{\text{cfm}} \left(1.08 \; \frac{\text{Btu-min}}{\text{ft}^3\text{-hr-}^\circ\text{F}} \right) (T_{id} - T_{od})
$$

$$
= \frac{\left(120 \; \dfrac{\text{ft}^3}{\text{hr}} \right) \left(1.08 \; \dfrac{\text{Btu-min}}{\text{ft}^3\text{-hr-}^\circ\text{F}} \right)}{60 \; \dfrac{\text{min}}{\text{hr}}}
$$

$$
= 118.8 \; \text{Btu/hr} \quad (120 \; \text{Btu/hr})
$$

The answer is B.

(j) From an appropriate source, the number of 65°F standard degree days for the location listed is 4811. The approximate heating value of #6 fuel oil is 150,000 Btu/gal. The annual savings per condominium is

$$
\text{savings} = \frac{\left(\dfrac{\text{cost}}{\text{gal}} \right) \left(24 \; \dfrac{\text{hr}}{\text{day}} \right) (\dot{q})(\text{DD})}{(T_i - T_o)(\text{HV}_{\text{Btu/gal}}) \eta_{\text{furnace}}}
$$

$$
= \frac{\left(\dfrac{\$1.50}{\text{gal}} \right) \left(24 \; \dfrac{\text{hr}}{\text{day}} \right) \left(200 \; \dfrac{\text{Btu}}{\text{hr-window}} \right)}{(70^\circ\text{F} - 15^\circ\text{F}) \left(150,000 \; \dfrac{\text{Btu}}{\text{gal}} \right) (0.55)}
$$

$$
= \$76.34 \quad (\$80 \; \text{per year})
$$

This answer assumes a 65°F basis for winter degree days. Internal heat gains in modern construction are able to maintain comfortable living conditions at temperatures slightly less than 65°F, and correction (reduction) factors to the degree days could be applicable if the upgraded condominiums qualify as "modern."

The answer is D.

18 SOLUTION

(a) The normal stress from beam bending is due to the applied eccentric load.

$$
\sigma = \frac{F}{A} \pm \frac{Mc}{I} = \frac{F}{wt} \pm \frac{Fe \left(\dfrac{t}{2} \right)}{\left(\dfrac{1}{12} \right) wt^3} = \frac{F}{wt} \pm \frac{6Fe}{wt^2}
$$

$$
= \frac{1.5 \; \text{lbf}}{(0.8 \; \text{in})(0.04 \; \text{in})} \pm \frac{(6)(1.5 \; \text{lbf})(0.6 \; \text{in})}{(0.8 \; \text{in})(0.04 \; \text{in})^2}
$$

$$
= 47 \; \text{psi} \pm 4219 \; \text{psia}
$$

$$
\sigma_{\max} = 47 \; \text{psi} + 4219 \; \text{psi}
$$

$$
= 4266 \; \text{psi} \quad (4270 \; \text{psi tension})
$$

The answer is C.

(b) $\quad \delta_x = \dfrac{FL^3}{3EI} = \dfrac{\dfrac{FL^3}{3E}}{\left(\dfrac{1}{12} \right) wt^2} = \dfrac{4FL^3}{Ewt^3}$

$$
= \frac{(4)(2 \; \text{lbf})(0.6 \; \text{in})^3}{\left(29 \times 10^6 \; \dfrac{\text{lbf}}{\text{in}^2} \right) (0.16)(0.040 \; \text{in})^3}
$$

$$
= 0.0058 \; \text{in}
$$

The answer is D.

(c) Check to see if the 0.8 in portion is a wide beam.

$$\frac{w}{t} = \frac{0.8 \text{ in}}{0.040 \text{ in}} = 20$$

This is greater than 10, so the beam is wide. The deflection should be reduced by multiplying by $(1 - \nu^2)$.

The tip deflection due to a moment is

$$\delta_y = \left(\frac{ML^2}{2EI}\right)(1 - \nu^2) = \frac{FeL^2(1 - \nu^2)}{2E\left(\frac{1}{12}\right)wt^3}$$

$$= \frac{(1.75 \text{ lbf})(0.6 \text{ in})(2.0 \text{ in})^2(1 - (0.3)^2)}{(2)\left(29 \times 10^6 \dfrac{\text{lbf}}{\text{in}^2}\right)\left(\dfrac{1}{12}\right)(0.8 \text{ in})(0.040 \text{ in})^3}$$

$$= 0.0154 \text{ in} \quad (0.015 \text{ in})$$

The answer is C.

(d) Since this is a static loading, the stress concentration factor is not used.

The tensile stress at point B is

$$\sigma = \frac{Mc}{I} = \frac{Fe\left(\dfrac{t}{2}\right)}{\left(\dfrac{1}{12}\right)wt^3} = \frac{6Fe}{wt^2}$$

$$= \frac{(6)(2 \text{ lbf})(0.6 \text{ in})}{(0.16 \text{ in})(0.040 \text{ in})^2}$$

$$= 28{,}125 \text{ lbf/in}^3 \quad (28{,}000 \text{ psi})$$

Since there is no stress in any other direction to combine, this is the principal stress.

The answer is C.

(e) As in part (d), the normal stress at point C is

$$\sigma = \frac{6Fe}{wt^2} = \frac{(6)(2 \text{ lbf})(0.25 \text{ in})}{(0.16 \text{ in})(0.040 \text{ in})^2} = 11{,}719 \text{ lbf/in}^2$$

From combined stress theory, the maximum shear stress is

$$\tau_{\max} = \pm\tfrac{1}{2}\sqrt{(\sigma_x - \sigma_y)^2 + (2\tau)^2}$$

$$= \pm\tfrac{1}{2}\sqrt{\left(11{,}719 \dfrac{\text{lbf}}{\text{in}^2} - 0\right)^2 + 0}$$

$$= \pm 5859 \text{ lbf/in}^2 \quad (5900 \text{ psi})$$

The average direct shear is $F/A = 313 \text{ lbf/in}^2$. However, the direct shear is zero at the outer surface where this moment is largest.

The answer is D.

(f) The minimum stress is zero.

As in part (d), the maximum stress is

$$\sigma = \frac{6Fe}{wt^2} = \frac{(6)(1 \text{ lbf})(0.6 \text{ in})}{(0.16 \text{ in})(0.040 \text{ in})^2} = 14{,}063 \text{ lbf/in}^2$$

Since there are no other forces, this is the principal stress.

The alternating stress is one-half of the stress difference.

$$\sigma_{\text{alt}} = \tfrac{1}{2}(\sigma_{\max} - \sigma_{\min}) = \left(\tfrac{1}{2}\right)\left(14{,}063 \frac{\text{lbf}}{\text{in}^2} - 0\right)$$

$$= 7032 \text{ lbf/in}^2 \quad (7000 \text{ psi})$$

The answer is B.

(g) From part (f), the mean stress is

$$\sigma_m = \tfrac{1}{2}(\sigma_{\min} + \sigma_{\max}) = \left(\tfrac{1}{2}\right)\left(0 + 14{,}063 \frac{\text{lbf}}{\text{in}^2}\right)$$

$$= 7032 \text{ lbf/in}^2 \quad (7000 \text{ psi})$$

The answer is C.

(h) The fatigue notch factor is calculated from the theoretical tensile stress concentration factor and the notch sensitivity factor.

$$q = \frac{K_f - 1}{K_t - 1}$$

$$K_f = 1 + q(K_t - 1) = 1 + (0.8)(1.4 - 1) = 1.32$$

The fatigue notch factor, as most stress concentration factors, is applied only to the alternating stress.

The factor of safety can be found graphically from a Goodman Line, or it can be calculated as

$$\text{FS} = \frac{S_e}{\sigma_{\text{eq}}} = \frac{S_e}{K_f\sigma_{\text{alt}} + \left(\dfrac{S_e}{S_{ut}}\right)\sigma_m}$$

$$= \frac{30{,}000 \dfrac{\text{lbf}}{\text{in}^2}}{(1.32)\left(12{,}000 \dfrac{\text{lbf}}{\text{in}^2}\right) + \left(\dfrac{30{,}000 \dfrac{\text{lbf}}{\text{in}^2}}{75{,}000 \dfrac{\text{lbf}}{\text{in}^2}}\right)\left(35{,}000 \dfrac{\text{lbf}}{\text{in}^2}\right)}$$

$$= 1.01 \quad (1.0)$$

The answer is B.

(i) $\sigma_{\max} = 32{,}000 \; \dfrac{\text{lbf}}{\text{in}^2} + 21{,}094 \; \dfrac{\text{lbf}}{\text{in}^2} = 53{,}094 \; \text{lbf/in}^2$

$\sigma_{\min} = 32{,}000 \; \dfrac{\text{lbf}}{\text{in}^2} + 3515 \; \dfrac{\text{lbf}}{\text{in}^2} = 35{,}515 \; \text{lbf/in}^2$

The alternating and mean stresses are

$\sigma_{\text{alt}} = \frac{1}{2}(\sigma_{\max} - \sigma_{\min})$

$= \left(\frac{1}{2}\right)\left(53{,}094 \; \dfrac{\text{lbf}}{\text{in}^2} - 35{,}515 \; \dfrac{\text{lbf}}{\text{in}^2}\right)$

$= 8790 \; \text{lbf/in}^2 \quad (8800 \; \text{psi tension})$

$\sigma_m = \frac{1}{2}(\sigma_{\min} + \sigma_{\max})$

$= \left(\frac{1}{2}\right)\left(35{,}515 \; \dfrac{\text{lbf}}{\text{in}^2} + 53{,}094 \; \dfrac{\text{lbf}}{\text{in}^2}\right)$

$= 44{,}305 \; \text{lbf/in}^2 \quad (44{,}000 \; \text{psi tension})$

The answer is B.

(j) The population standard deviation, σ, is

$\sigma = \sqrt{\text{variance}} = \sqrt{5760 \times 10^3 \; \dfrac{\text{lbf}^2}{\text{in}^4}} = 2400 \; \text{lbf/in}^2$

The standard normal variable is

$z = \dfrac{x_0 - \mu}{\sigma} = \dfrac{23{,}250 \; \dfrac{\text{lbf}}{\text{in}^2} - 30{,}000 \; \dfrac{\text{lbf}}{\text{in}^2}}{2400 \; \dfrac{\text{lbf}}{\text{in}^2}} = -2.81$

From a normal table, the area under the normal curve between 0 and 2.81 is 0.4975. The probability that the value will be less than 2.81 is

$p\{z < 2.81\} = 0.5 + 0.4975 = 0.9975 \quad (99.8\%)$

The answer is D.

19 SOLUTION

(a) $180°$ is π radians.

$\dfrac{T_1}{T_2} = e^{f\phi}$

$f = \dfrac{\ln\left(\dfrac{T_1}{T_2}\right)}{\phi} = \dfrac{\ln\left(\dfrac{5750 \; \text{lbf}}{3250 \; \text{lbf}}\right)}{\pi}$

$= 0.1816 \quad (0.18)$

The answer is B.

(b) Sum the moments from forces in the x-direction about point D.

$\displaystyle\sum_{x\text{-direction}} M_{\text{D}} = (5750 \; \text{lbf} + 3250 \; \text{lbf})(5 \; \text{in}) - A_x(20 \; \text{in})$

$= 0$

$A_x = 2250 \; \text{lbf}$

D_x carries the remainder of the forces in the x-direction.

The answer is C.

(c) Find the reaction at A in the z-direction by summing moments from forces in the z-direction about point D.

$\displaystyle\sum_{z\text{-direction}} M_{\text{D}} = (4000 \; \text{lbf} + 2000 \; \text{lbf})(15 \; \text{in}) - A_z(20 \; \text{in})$

$= 0$

$A_z = 4500 \; \text{lbf}$

Sum the moments from forces in the z-direction from point E and bearing A.

$M_x = \sum M_{\text{E-A}}$

$= (4000 \; \text{lbf} + 2000 \; \text{lbf})(5 \; \text{in}) - (4500 \; \text{lbf})(10 \; \text{in})$

$= -15{,}000 \; \text{in-lbf}$

(The sign will depend on the direction chosen for positive moments.)

The answer is A.

(d) Work with the pulley at point B.

$\text{torque} = M_y = F_{\text{net}} r$

$= (4000 \; \text{lbf} - 2000 \; \text{lbf})\left(\dfrac{15 \; \text{in}}{2}\right)$

$= 15{,}000 \; \text{in-lbf}$

(The symbol M_{xz} is also used for this torque.)

The answer is B.

(e) Sum the forces in the z-direction from point E to bearing A.

$V_z = \displaystyle\sum_{z\text{-direction}} F_{\text{E-A}}$

$= -4000 \; \text{lbf} - 2000 \; \text{lbf} + 4500 \; \text{lbf}$

$= -1500 \; \text{lbf}$

The answer is B.

(f) The area moment of inertia of the circular shaft is

$$I = \left(\frac{\pi}{4}\right) r^4 = \left(\frac{\pi}{4}\right) \left(\frac{3 \text{ in}}{2}\right)^4 = 3.976 \text{ in}^4$$

Normal stress at this point is in the y-direction and is caused by forces in the z-direction (i.e., moments about the x-axis).

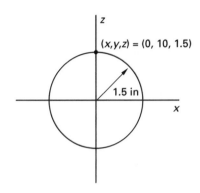

Use the new, given data.

$$\sigma_y = \frac{M_x c}{I} = \frac{(22,500 \text{ in-lbf}) \left(\frac{3 \text{ in}}{2}\right)}{3.976 \text{ in}^4}$$
$$= 8488 \text{ lbf/in}^2 \quad (8500 \text{ psi})$$

The answer is D.

(g) For a circular shaft, the maximum shear stress is

$$\tau_{\text{max},z} = \frac{4V_z}{3A} = \frac{(4)(5000 \text{ lbf})}{(3) \left(\frac{\pi}{4}\right) (3 \text{ in})^2}$$
$$= 943 \text{ lbf/in}^2 \quad (940 \text{ psi})$$

The answer is D.

(h) The polar moment of inertia is

$$J = \left(\frac{\pi}{2}\right) r^4 = \left(\frac{\pi}{2}\right) \left(\frac{3 \text{ in}}{2}\right)^4 = 7.952 \text{ in}^4$$

$$\tau_y = \frac{M_y c}{J} = \frac{(25,000 \text{ in-lbf}) \left(\frac{3 \text{ in}}{2}\right)}{7.952 \text{ in}^4}$$
$$= 4716 \text{ lbf/in}^2 \quad (4700 \text{ psi})$$

(The symbol τ_{xz} is also used for this shear stress.)

The answer is D.

(i) Direct shear stress is zero at the outer surface of the shaft.

At the point identified, $\sigma_z = 0$ because it is in the neutral plane of bending.

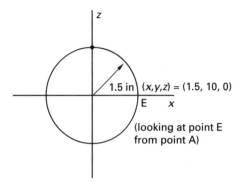

Normal stress at the point is in the y-direction and is caused by forces in the x-direction (i.e., moments about the z-axis).

$$\sigma_y = \frac{M_z c}{I} = \frac{(20,000 \text{ in-lbf}) \left(\frac{3 \text{ in}}{2}\right)}{3.976 \text{ in}^4} = 7545 \text{ lbf/in}^2$$

From combined stress theory, the maximum shear stress is

$$\tau_{\text{max}} = \pm \tfrac{1}{2} \sqrt{(\sigma_x - \sigma_z)^2 + (2\tau_y)^2}$$
$$= \pm \tfrac{1}{2} \sqrt{\left(7545 \frac{\text{lbf}}{\text{in}^2} - 0\right)^2 + \left[(2) \left(4716 \frac{\text{lbf}}{\text{in}^2}\right)\right]^2}$$
$$= \pm 6039 \text{ lbf/in}^2 \quad (6000 \text{ psi})$$

The answer is D.

(j) For combined stress theory, the maximum normal stress is

$$\sigma_{\text{max}} = \left(\tfrac{1}{2}\right) (\sigma_y + \sigma_z) \pm \tau_{\text{max}}$$
$$= \left(\tfrac{1}{2}\right) \left(7545 \frac{\text{lbf}}{\text{in}^2} + 0\right) \pm 6039 \frac{\text{lbf}}{\text{in}^2}$$
$$= 9812 \text{ lbf/in}^2 \quad (9800 \text{ psi})$$

The answer is D.

20 SOLUTION

(a) The appropriate formula for the natural frequency (in English units) is

$$\omega = \sqrt{\frac{kg_c}{I}}$$

The density and diameter of the steel balls are needed to solve for the mass of the steel balls. This mass and the distance from the rotational axis determines the mass moment of inertia. The spring constant and g_c are then used to find the natural frequency.

The answer is C.

(b)
$$\theta = \frac{T}{k} = \frac{0.090 \text{ in-lbf}}{2.857 \times 10^{-3} \dfrac{\text{in-lbf}}{\text{rad}}} = 31.5 \text{ rad}$$

Convert this to revolutions.
$$\frac{31.5 \text{ rad}}{2\pi \dfrac{\text{rad}}{\text{rev}}} = 5.01 \text{ rev} \quad (5 \text{ rev})$$

The answer is D.

(c) This is an example of undamped free vibration. An argument could be made that it is damped by friction and then forced by the battery's energy, but there is no choice for damped forced vibration.

The answer is B.

(d) Since (from part A) the natural frequency is proportional to the square root of the spring constant,
$$\sqrt{2} = 1.41 \quad \text{[i.e., 41\% increase]}$$

The answer is C.

(e) The gravitational constant has no effect on rotational systems. The g_c term in the equation converts from pounds to slugs. It is not the local gravitational acceleration.

The answer is D.

(f) The rotational angle is
$$\theta = \frac{(720°)\left(2\pi \dfrac{\text{rad}}{\text{rev}}\right)}{\dfrac{360°}{\text{rev}}} = 12.57 \text{ rad}$$

$$T = k\theta = \left(2.857 \times 10^{-3} \frac{\text{in-lbf}}{\text{rad}}\right)(12.57 \text{ rad})$$
$$= 0.0359 \text{ in-lbf} \quad (0.036 \text{ in-lbf})$$

The answer is B.

(g) Increasing dimension B will increase the mass moment of inertia. To compensate, the mass of the balls must be reduced.

The answer is D.

(h) The natural frequency of a system remains essentially unchanged by damping. The amplitude will decrease asymptotically to zero.

The answer is B.

(i) Friction provides a damping force, but there is no forcing function.

The answer is B.

(j) The mass of the three balls is
$$m = 3\rho V = 3\rho \left(\tfrac{4}{3}\right)\pi r^3$$
$$= (3)\left(0.284 \frac{\text{lbm}}{\text{in}^3}\right)\left(\tfrac{4}{3}\right)\pi(0.200 \text{ in})^3$$
$$= 0.02855 \text{ lbm}$$

The mass moment of inertia of a sphere about its centroidal axis is
$$I_c = \left(\tfrac{2}{5}\right)mr^2$$
$$- \left(\tfrac{2}{5}\right)(0.02855 \text{ lbm})(0.200 \text{ in})^2$$
$$= 0.0004568 \text{ lbm-in}^2$$

Use the parallel axis theorem.
$$I = I_c + mr^2$$
$$= 0.0004568 \text{ lbm-in}^2 + (0.02855 \text{ lbm})(0.75 \text{ in})^2$$
$$= 0.0165 \text{ lbm-in}^2$$

(Notice that the centroidal moment of inertia did not contribute much to this value.)

The natural frequency is
$$\omega = \frac{1}{2\pi}\sqrt{\frac{kg_c}{I}}$$

$$= \left(\frac{1}{2\pi}\right)\sqrt{\frac{\left(2.875 \times 10^{-3} \dfrac{\text{in-lbf}}{\text{rad}}\right) \times \left(32.2 \dfrac{\text{ft-lbm}}{\text{lbm-sec}^2}\right)\left(12 \dfrac{\text{in}}{\text{ft}}\right)}{0.0165 \text{ lbm-in}^2}}$$
$$= 1.306 \text{ Hz} \quad (1.3 \text{ Hz})$$

The answer is A.

2848